UP-TO-SPEED **MATH**

BOOK **2**

STORY PROBLEMS

by Andrew M. Schor
Dawn Purney
Thomas H. Hatch

UP-TO-SPEED MATH

Number Sense

Operations

Working with Fractions, Decimals, and Percents

Geometry and Measurement

Patterns, Functions, and Algebra

Data Analysis and Reasoning

Math Standards Review 1

Math Standards Review 2

Story Problems Book 1

Story Problems Book 2

Story Problems Book 3

created by **Kent Publishing Services, Inc.**
designed by **Signature Design Group, Inc.**

SADDLEBACK PUBLISHING, INC.
Three Watson
Irvine, CA 92618-2767

Website: www.sdlback.com

ISBN 1-56254-522-1

Printed in the United States of America

09 08 07 06 05 6 5 4 3 2 1

TABLE OF CONTENTS

TABLE OF CONTENTS

TO THE STUDENT

This book contains a sequence of lessons designed to help you master the art of solving word and story problems. The lessons are organized into two sections. In Part 1, you'll learn the four-step process for problem solving. The logo at the upper left corner of each lesson will remind you to keep thinking about this process as you work.

In the lessons in Part 2, you'll learn and practice specific math skills and problem solving strategies, such as:

- Act it out
- Draw a picture or diagram
- Guess and check
- Look for a pattern
- Make a graph
- Make a list
- Make a table
- Use proportions
- Work a simpler problem
- Work backwards

Story Problems lessons are two page spreads with instruction and worked out examples in the upper part of the left hand page. The exercises that follow the instruction provide guided and independent practice. So, get up-to-speed, and look forward to becoming a master problem solver.

SKIMMING AND SCANNING

To solve a story problem, you need to first understand the question. Scan for key words or familiar patterns. Skim for main ideas to help determine the question.

Joseph had 20 pencils. His friend Mya had three-quarters as many. How many pencils did Mya have?

Key phrases to note are:

20 pencils
three-quarters as many
how many pencils

Some word problems do not use a question mark to signal the question.

Joseph had 20 pencils. His friend Mya had three-quarters as many. Find the number of pencils Mya had.

The question is the same as before. It could also be stated "How many is 3/4 of 20 pencils?"

Read the problem thoroughly and check if the question you formed is what the question is really asking.

 A   **Which best states the question asked in the problem?**

1. During September, October, November, and December, Jeff worked after school on 32 days. He worked the same number of days each month. Find the number of days Jeff worked each month.

   Ⓐ How many days did Jeff work during October?

   Ⓑ How many days did Jeff work each month?

   Ⓒ Did Jeff work the same number of days each month?

   Ⓓ Will Jeff work for 32 days the next four months?

October is not the only month mentioned, so A is wrong. The question states that Jeff worked the same amount each month, so C is incorrect. There is no mention of the future, so D is incorrect. Choice B correctly states the question. Now try the next one on your own.

2. Vanessa ran 36 miles the first week, 31 miles the second week, and 28 miles the third week. About how many miles did she run in all?

Ⓐ Compare the number of miles she ran the first week compared to the third week.

Ⓑ Predict the number of miles Vanessa will run during the next 3 weeks.

Ⓒ Estimate the total number of miles Vanessa ran the first 2 weeks.

Ⓓ Estimate the total number of miles Vanessa ran during the 3 weeks.

B State in another way the question being asked in the problem.

1. Lauren, Loretta, Andy, and Sean sat together in the first row of the auditorium. Lauren sat next to Sean. Sean sat next to Loretta, who was in the fourth seat. Andy was in the first seat at the end of the row. Who sat in the second seat in the row?

2. Lake Tahoe is on the border of California and Nevada. This beautiful body of water is 22 miles long and 12 miles wide. If the lake is shaped roughly like a rectangle, what is the approximate distance around it?

C Solve. Show your work on another piece of paper.

1. Leon and Sue bought 7 packages of construction paper for their art project. Each package of paper cost $3. Leon contributed $9 to the total cost. How many packages of construction paper did Sue pay for?

2. Many people use pickup trucks to tow their trailers. Ken is comparing two pickup trucks. The salesperson tells him the first truck can tow up to 8,100 pounds, while another truck can pull up to 2,200 pounds more. Find how many pounds the second truck can tow.

3. The Smiths are putting a shed in their backyard to store tools and their lawnmower. At the garden center, sheds are being sold for $992. A special 6-foot door can be added for an additional $99. They have $1,050 to spend. Can they afford the special door?

L E S S O N   2

RECOGNIZING MISSING INFORMATION

Sometimes you may not be able to solve a story problem because it does not give you all the information you need. After you determine what question the problem is asking, reread the problem to see if all the necessary information is provided.

Toshi spent 2 quarters, 1 nickel, and 1 dime. How much money did he have left?

You can easily find how much money Toshi spent, but, to answer the question asked, you need to know how much money Toshi started with.

 A   Choose the best answer.

1. Which can be solved?

   Ⓐ One bottle holds 0.50 liters of soda and another holds 1.75 liters. How many liters of soda are in the two bottles combined?

   Ⓑ One bottle holds 0.50 liters of soda and another bottle holds more. How many liters of soda are in the two bottles combined?

   Ⓒ One bottle holds 0.50 liters of soda and another bottle holds 0.50. How many liters of soda did they drink?

   Ⓓ A bottle of soda is smaller than a second bottle holding 1.75 liters. How many liters of soda are in the two bottles combined?

   Choice A is correct. It is the only one you can solve. Now try this one.

2. Which has all the information you need to tell what time Ms. Green's class left the museum?

   Ⓐ Ms. Green's class stayed at the museum for 3 hours and 20 minutes.

   Ⓑ Ms. Green's class arrived at the museum at 9:15 A.M. and left in the afternoon.

   Ⓒ Ms. Green's class arrived in the morning and left 3 hours and 20 minutes later.

   Ⓓ Ms. Green's class arrived at 9:15 A.M. and left 3 hours and 20 minutes later.

B Write a phrase or sentence to tell what information is needed to solve the problem.

1. In a basketball playoff game, the Indiana Pacers beat the New Jersey Nets. The star of the Pacers was Jermaine O'Neal, who scored 30 points. Find how many points his teammates scored.

2. In 2000, Janine's salary was raised to $35,000 a year. She expects the same raise each year after that. How much can Janine expect to earn in 2003?

3. Riana arrived at the park at 3:30; Halley came later. They left together 2 hours after that. When did they leave?

4. Ricardo has 5 different pairs of socks and some pants to choose from. How many combinations of pants and socks can Ricardo put together?

C Solve, if possible. If not, state the information needed.

1. Cosmos are beautiful flowers that are easy to grow. Once the seeds are planted, they take about 12 days to sprout. Karena planted her seeds on June 15. On what date might she expect them to sprout?

2. The senior class wants to rent vans for a class trip. Each van holds 8 students. How many vans should they rent?

3. In the spring of 2002, there was a drought in the northeast United States. Rainfall measured 2.66 inches below normal for New York City for January through April. What was the normal rainfall for those 4 months?

4. In the first 3 hours of sales, 300 tickets were sold to the new movie. In the next 3 hours, twice as many tickets were sold. How many tickets were sold in those 6 hours?

L E S S O N   3

SELECTING IMPORTANT INFORMATION

After you identify the question and determine whether you can solve it, find the facts you need to solve the problem. Sometimes a story problem has extra information that may be interesting, but will not help you solve the problem.

Topaz is 12, and she enjoys painting. Last year she painted 16 paintings and this year she painted 13 paintings. How many pictures did she paint during the two years?

What information is needed to solve this problem?

• Topaz painted 16 pictures last year.

• Topaz painted 13 pictures this year.

Knowing that Topaz is 12 will not help you solve the problem.

Make sure you have all the information you need before trying to solve a story problem.

A **What facts do you need to solve the problem?**

1. There are 3 classes in the fifth grade: 5-1, 5-2, and 5-3. Class 5-2 is the largest with 28 students. If each class averages 26 students, how many students are in fifth grade?

   Ⓐ Each class in the fifth grade averages 26 students.

   Ⓑ The fifth grade classes are 5-1, 5-2, and 5-3

   Ⓒ There are 3 classes, and each averages 26 students.

   Ⓓ Class 5-2 is the largest class with 28 students

   Although you need the information in A, you also need another piece of information. Both B and D may be interesting, but they won't help you solve the problem. Choice C has the information you will need to solve the problem. Do the rest on your own.

2. There were 65 riders entered in last week's bike race. The race was 24 miles, and was expected to last about 2 hours. If 21 riders did not finish the race, how many riders completed the full distance?

   Ⓐ There were 65 riders, and 21 did not finish.

   Ⓑ The bike race covered a distance of 24 miles.

   Ⓒ The bike race will last about 2 hours.

   Ⓓ There were 65 riders entered in the bike race.

B List the facts needed to solve the problem.

1. Ross drew a diagram of his bedroom, which has an 8-foot ceiling and a desk with two chairs. If he used a scale of 1 inch = 2 feet, what will the length of his drawing be if the bedroom is 14 feet in length?

2. On Sunday night, there is a movie on Channel 7. There is also a comedy on Channel 5 lasting 30 minutes. If the movie starts at 8:00 P.M. and ends at 10:30 P.M. how long is the movie?

3. Jillian read a 300-page biography for her book report on 5 nights last week. She read 25 pages each night. If she reads at the same rate, how many pages could she read in 10 days?

C Solve. Use the chart to answer the questions.

State	Area (square miles)	Highest Elevation (feet)	Lowest Elevation (feet)	Date of Statehood
West Virginia	24,231	4,863	240	1863
North Dakota	70,702	3,506	750	1889
Illinois	56,345	1,235	279	1818
Nevada	110,561	13,140	470	1864

1. Kansas has an area greater than North Dakota, but less than Nevada. Lou says that the area of Kansas is less than 70,000 square miles. Is he correct?

2. Which state in the table has the greatest difference in elevation between the highest and lowest points?

CHOOSING AN OPERATION

Once you have all the important information, you can plan how to solve a word problem. Understanding the question will help. Also, signal words can be used as clues to help decide which operations to use.

For the sixth-grade fundraiser, Ms. Chang's class raised $430, Mr. Brooks's class raised $395, and Mrs. Ketch's class raised $420. What is the total amount the sixth-graders raised?

The words "total amount" usually signal addition. Since you want to find the total amount of money raised, you will add. The classes earned $430 + $395 + $420 = $1,245 in total.

Sometimes you will need to use more than one operation to solve a story problem.

There are 11 players on a soccer team. 4 of the players are defenders, 2 are forwards, and 1 is a goalie. The rest are midfielders. How many players are midfielders?

You need to find the total number of midfielders on a soccer team.

One way to solve this problem is to add the number of players at other positions.

4 + 2 + 1 = 7 players.

Then, subtract that number from 11.

11 - 7 = 4 players are midfielders.

A Which operation will solve the problem?

1. Each day last week, Mulan played baseball for 2 hours. How many hours did she play baseball last week altogether?

   Ⓐ add

   Ⓑ subtract

   Ⓒ multiply

   Ⓓ divide

You could solve the problem using repeated addition (2+2+2+2+2+2+2=14), but multiplying is easier. Choice C is correct—multiply 7 days by 2 hours. Determine the operation for next problem on your own.

2. Tefo spent $4.50 for a sandwich and $2.00 for a drink. If he paid with a $10 bill, how much change did he receive?

   Ⓐ multiply and subtract   Ⓒ add and subtract

   Ⓑ add and divide   Ⓓ divide and add

 B   Write a phrase or sentence telling how to solve the problem.

1. Brian is digging a garden in his backyard. The length will be 8 feet and the width will be 6 feet. If he wants to put a fence around the garden, how many feet of fencing should he buy?

2. There are 400 fans in the stands at a baseball game. If there are 4 equal sections in the stands, how many fans are in each section?

3. It is very important that we all try to save water. Putting in a new toilet that uses 1.6 gallons per flush will help. New showerheads that can't use more than 2.5 gallons of water per minute will also help conserve water. If a toilet is flushed 5 times a day and a shower runs for 6 minutes daily, how many of gallons of water will be used?

C   Solve. Show your work on another sheet of paper.

1. During the California gold rush of 1849, supplies were very expensive. Things cost 10 times as much as they did back East. If a shovel cost $0.80 on the East coast, how much would it have cost in California?

2. The monarch butterfly is an amazing creature. This tiny insect can migrate south 1,000 miles each year in the fall. In the spring, they return north. How many miles might a monarch butterfly migrate each year?

3. So far, Mariah has used 2 rolls of film on her vacation. There are 24 pictures on each roll of film. If she had enough film for 72 pictures, how many rolls of film did Mariah take with her?

WRITING A NUMBER SENTENCE

When solving a story problem, use what you know about finding important information and choosing operations to write a number sentence.

A 385-mile trip by car took 7 hours. What was the average speed in miles per hour?

Number Sentence
$385 \div 7 = ?$

Now you can more easily solve the problem.
$385 \div 7 = 55$

Answer
55 miles per hour

A Choose the number sentence to solve the story problem.

1. Samuel had 15 pens. He gave 3 to Tamara and then bought 4 more. How many pens does Samuel have now?

   Ⓐ $4 + 3 - 15 = ?$ Ⓒ $15 - 3 + 4 = ?$

   Ⓑ $15 + 4 + 3 = ?$ Ⓓ $3 + 15 - 4 = ?$

   Samuel started with 15 pens. He lost 4 pens, but gained 3. C is correct. Now try these.

2. Trina had 8 boxes of cookies with 16 cookies in each box for the school bake sale. How many cookies are for sale?

   Ⓐ $16 \times 8 = ?$ Ⓒ $16 \div 8 = ?$

   Ⓑ $16 - 8 = ?$ Ⓓ $16 + 8 = ?$

3. Li is 4 years younger than twice her son's age. Her son is 20. How old is Li?

   Ⓐ $4 + (20 \div 2) = ?$ Ⓒ $2(20 + 4) = ?$

   Ⓑ $(20 \times 2) - 4 = ?$ Ⓓ $(20 \times 2) + 4 = ?$

B **Will the number sentence solve the problem? Explain your answer.**

1. Cliff had 297 trading cards and Felicia had 291 cards in their collections. Cliff sold 20 cards and Felicia bought 18 cards. How many cards do they have altogether?

    Number Sentence: 297 − 20 + 291 + 18 = ?

2. Parvati owns 8 pairs of socks. How many socks does he have?

    Number Sentence: 8 ÷ 2 = ?

3. A touchdown in football is worth 6 points. Kenton scored 5 touchdowns. How many points did he score?

    Number Sentence: 5 x 6 = ?

C **Write a number sentence, then solve. Show your work on another piece of paper.**

1. Apples cost $1 per pound. Oranges cost $0.50 each. Find the cost of 3 pounds of apples and 6 oranges.

2. From the 1940s through 1990, steel production first rose, then fell. In 1940, about 60 million tons of steel was produced. This amount increased by 65 million tons by 1970. In the next 20 years, production decreased by 35 million tons. What was steel production in 1990?

3. Many doctors believe the more you laugh, the better you will feel. In one study, it was found that most children laugh about 400 times a day. Adults could use more laughs, since they only laugh 15 times daily. What is the difference in the number of times a child laughs in a week compared to the average adult?

4. Malik is 2 inches shorter than Tony. Tony is 5 inches taller than Tom. If Tom is 62 inches tall, how tall is Malik?

WRITING AN EQUATION

An equation shows that two amounts are equal. A letter is often used to hold the place of the unknown number.

$$74 \times 8 = x$$

$$15 \div y = 2.5$$

$$n - 57 = 134$$

If normal yearly rainfall in New York City is 47.25 inches and 37.73 inches has fallen this year, how far below normal is this year's rainfall?

Equation

$$47.25 - 37.73 = r$$

Answer

$$r = 9.52 \text{ inches}$$

Elly and Whitley wanted to train together for a bike race. On Sunday they biked 27 miles; on Tuesday, 26 miles; and on Thursday, 32 miles. How many miles did the two bikers ride altogether that week?

Equation

$$2 (27 + 26 + 32) = m$$

$$2 (85) = m$$

Answer

$$m = 170 \text{ miles}$$

A Choose the equation to use for the story problem.

1. Vince had $5.00 to spend on snacks. He bought chips for $2.95 and pretzels for $2.00. When he arrived home, his mother gave him another $2.00. How much money did Vince now have?

    Ⓐ $5.00 - $2.95 − $2.00 + $2.00 = n

    Ⓑ $2.95 + $2.00 − $2.00 = n

    Ⓒ $5.00 − $2.95− $2.00 = n

    Ⓓ $5.00 + $2.95 + $2.00 = n

    Only A is complete. Every other choice leaves out something. Now look at the next problem.

2. Margie had 3 boxes of books with 8 books in each box. If she gave 6 books to her friend, Dominique, how many books does Margie have left?

    Ⓐ $(3 \times 8) \times 6 = t$

    Ⓑ $(3 \times 8) + 6 = t$

    Ⓒ $(3 \times 6) - 8 = t$

    Ⓓ $(3 \times 8) - 6 = t$

Write an equation for the problem.

1. On Monday, it rained 0.01 inches in Jackson, Mississippi. Pittsburgh, Pennsylvania received 20 times as much rainfall as Jackson. What was the combined rainfall of the two locations?

2. Ken created 28 greeting cards. He kept 4 to use later and sent an equal number of the rest of the cards to 6 relatives. How many cards did Ken send to each relative?

3. Each student in Mr. Moore's class is expected to read 35 books this year. So far, Amanda has read 15 books. Stuart has read 1/3 as many as Amanda. How many books has Stuart read so far?

 Write an equation and solve. Show your work on another sheet of paper.

The Presidential Mountain Range in New Hampshire

Mountain	Elevation
Mount Madison	5,367 feet
Mount Clay	5,533 feet
Mount Washington	6,288 feet
Mount Eisenhower	4,780 feet

1. Julia climbed Mount Madison, Mount Clay, and Mount Eisenhower last summer. What was the combined elevation of the mountains she climbed last summer?

2. When Julia climbed Mt. Eisenhower, she stopped to rest 1/2 way up. Find the elevation of the place where she stopped.

3. Mount Monroe is 904 lower than Mount Washington, but 383 feet taller than Mount Franklin. What is the elevation of Mount Franklin?

REVIEW

Once you've gathered information and made a plan, you can calculate a solution.

A group of 5 students raised $380 selling used books at the library fundraiser last week. If each raised the same amount, how much did each student raise?

What we know Five students raised $380 together. They each raised the same amount.

Our plan Write an equation using division to solve this problem.

Calculate Let m = money each student raised.

Then, $\$380 \div 5 = m$

$m = \$76$.

Each student raised $76 for the fundraiser.

 A Choose the best answer.

28 adults are sitting in a movie theater. There are 6 more children than adults. How many people are in the theater?

1. Which states the information needed to solve the problem?

   Ⓐ There are 28 people sitting in the movie theater.

   Ⓑ There are 6 more children than adults in the theater.

   Ⓒ There are 28 adults and 6 more children than adults.

   Ⓓ There are 28 adults and 6 children sitting in the theater.

   The problem tells us the number of adults. It also tells you how many more children there are than adults. Choices A and B do not include enough information. Also, D is incorrect, since the problem does not state that there are 6 children. Choice C states all the key facts needed.

2. Which number sentence can be used to solve this problem?
   Ⓐ 28 + 6 + 28 = ?   Ⓑ 28 + 6 + 22 = ?   Ⓒ 28 − 6 + 28 = ?   Ⓓ 28 + 28 − 6 = ?

3. How many people are in the theater?

Ⓐ 34   Ⓑ 68   Ⓒ 56   Ⓓ 62

B List the facts you need, then describe your plan.

1. There are 700 cars in a mall parking lot. 130 of the cars are blue and 160 are black. How many of the cars in the parking lot are neither blue nor black?

2. Beginning in the 1500s, the Spanish sent many people to explore what is now the southwest United States. In 1539, Marcos de Niza set out to find gold. The following year, Francisco Vasquez de Coronado began his journey. Other explorers followed, and 70 years later the Spanish founded the city of Santa Fe. When was Santa Fe founded?

C Solve. Show your work on another piece of paper.

1. During the three-day art show, 2,037 people attended. On the first day of the art show, 791 people visited. On the second day, 653 people saw the exhibits. How many people went to the art show on the third day?

2. In 2000, there were 650 students in the Main Street School. Another 46 students enrolled in the school each of the next 3 years. How many students were in the school in 2004?

3. In their first 5 home games, the New York Yankees averaged 38,679 fans. In the next 6 home games, the average attendance rose to 41,154 because of the warmer weather. What was the total attendance for the first 11 home games?

4. In a survey at the mall, it was found that one-fourth of the people questioned were shopping for shoes. If 68 people said they were shopping for shoes, how many people were surveyed?

L E S S O N 2

CALCULATING SOLUTIONS

When adding or subtracting, make sure to line up the numbers at the ones place, or on the decimal point.

And to check, use *inverse operations*.

$$\begin{array}{r} 38.7 \\ +\ 6.23 \\ \hline 44.93 \end{array} \qquad \begin{array}{r} 44.93 \\ -\ 6.23 \\ \hline 38.70 \end{array}$$

Some problems involve *regrouping*.

$$\begin{array}{r} \scriptstyle 5\ \ 1211\ 11 \\ \cancel{6,321} \\ -\ 2,354 \\ \hline 3,967 \end{array} \qquad \begin{array}{r} \scriptstyle 1 \\ \$1.64 \\ +\ \$0.75 \\ \hline \$2.39 \end{array}$$

A *factor* is a number that divides evenly into another number. For example, 2 is a factor of 8. Also, notice that 8 may be evenly divided by 2.

Most numbers have more than two factors. The factors of 12 are 1, 2, 3, 4, 6, and 12.

Multiples are products of two numbers.

The multiples of 3 are 3, 6, 9, 12, 15, 18, …

The multiples of 4 are 4, 8, 12, 24, 28, …

  A Choose the letter of the correct answer.

1. 12,197 + 45 + 798 + 4,003 =

Ⓐ 58,532   Ⓑ 60,657   Ⓒ 16,823   Ⓓ 17,043

Compare your calculations with the problem below to see that D is correct. Then, do the next problem on your own.

$$\begin{array}{r} \scriptstyle 1\ \ \ 2\ 2 \\ 12,197 \\ 45 \\ 798 \\ +\ 4,003 \\ \hline 17,043 \end{array}$$

2. 731,426 − 58,738 =

   Ⓐ 672,688   Ⓑ 683,798   Ⓒ 655,009   Ⓓ 681,438

 B Write a phrase or sentence describing how you will solve the problem.

1. Christopher is putting 64 cookies in boxes to give as gifts. If he puts an equal number of cookies in each box, can he give a box to each of 12 friends and have no cookies left over?

2. Diana is planning her flower garden. One row will be planted with nasturtiums, which are beautiful red flowers on a low vine. The second row will be planted with marigolds. If the nasturtiums are planted 2 inches apart, and the marigolds are planted 3 inches apart, how many plants of each variety will be needed to fill out rows that are 36 inches long?

C Solve. Show your work on another piece of paper.

1. The oldest known stars in the universe are in a cluster known as M4, which was spotted in 2002 by the Hubble Space Telescope. These stars are estimated to be 12,700,000,000 years old. Scientists say that the M4 cluster may even be 500,000,000 years older than the current estimate! How old might M4 be?

2. Manuel bought 2 notebooks for $2 each, 2 sets of pens and pencils for $1.50 each, and an eraser for $.15. When he gave the cashier $22, she told him that it was too much money. Where did Manuel make a mistake in his calculations?

3. Alyssa's teacher likes to start every day with a math brainteaser. He lists two cities on the board: Boise, Idaho and Philadelphia, Pennsylvania. On a typical May day, their high temperatures had 6 as a greatest common factor. One city's high temperature was a multiple of 10. If the temperature in Philadelphia was less than 10 degrees higher than Boise, what might be the high temperature of each city? (Hint: first list multiples of 6. Then think about reasonable spring temperatures.)

1	Gather Information
2	Plan
3	Calculate
4	Check

L E S S O N 3

CHECKING SOLUTIONS

After you solve a problem, check your work by asking, "Does my solution answer the question?"

Mr. Sanford baked 72 rolls. Mrs. Po bought $\frac{1}{3}$ of all the rolls. How many more would she need to have 36 rolls?

Here's how LaRue solved the problem. Does her solution answer the question?

Let r = rolls Mrs. Po bought.

$72 \times \frac{1}{3} = r$

$24 = r$

Mrs. Po bought 24 rolls.

Although LaRue's answer is true, it does not answer the question. The question asks how many more rolls Mrs. Po needs to have 36. Answer: She needs 12.

A Choose the solution that answers the question.

1. Alex walks to the library 4 days each week. It is 2 miles to get there and 3 miles to get home since he takes the long way. How many miles does Alex walk to and from the library each week?

    Ⓐ 5 miles   Ⓑ 12 miles   Ⓒ 20 miles   Ⓓ 8 miles

Each day Alex walks 5 miles as he goes to the library and returns home. Choice A is incorrect since this is the mileage for one day. Choice B gives only the total going from the library, D only gives the total to the library. The answer is found by multiplying the daily mileage (5) by the number of days (4). Choice C is correct, since $5 \times 4 = 20$. Now try this problem.

2. Kristina had $50 to spend on school supplies. She bought 3 notebooks at $5 each, 2 packages of paper at $3 each, and some pens for $8. How much money did Kristina have left after she bought the supplies?

    Ⓐ $21   Ⓑ $29   Ⓒ $35   Ⓓ $15

B Write a sentence or two to explain your answer.

1. Mr. Lee wanted to park his car in a parking lot that charges $10 for the first hour and $3 for each additional hour. He wants to park there for 5 hours and determines it will cost him $22. Is he correct?

2. In September 2001, a rare Sumatran Tiger was born in the National Zoo in Washington, D.C. The cub, named Berani, is one of the few Sumatran tigers born in captivity. Worldwide there are about 500 of these animals living in the wild, and another 250 living in 85 zoos around the world. Karena figured that there were about 900 Sumatran tigers living in the wild and in zoos in all. Is she correct?

C Solve. Show all your work on another piece of paper.

Over the years, the Everglades in Florida has been drained. To create more dry land, over 1,400 miles of canals and dikes were built to drain water from the area. People became alarmed when they realized the harm this was causing. For example, the population of some birds has been reduced by 90%. Also, 68 species of animals are now either endangered or threatened.

As a solution, there is a plan to pump 80% of the water back into the Everglades. About 240 miles of canals and dikes will also be removed. It is hoped this will increase the amount of water available to plants and animals.

1. About 2 billion gallons of water flow out of the Everglades each day. According to the plan, about how many gallons will be pumped back into the area each week?

2. After the canals and dikes are removed as planned, about how many miles of these structures will remain in the Everglades?

3. The Everglades and the surrounding area cover about 18,000 square miles. If the Everglades is about 60 miles wide, what is its length? (The formula for determining area is $A = $ length x width.)

CHECK WITH ESTIMATION

When solving problems, we use estimation to:

- **predict the answer before solving**
- **check the solution**

Arturo wanted to read a 185-page novel in 5 days. He decided that he needed to read 20 pages each day to finish the book in time. Is he correct?

How can you tell if Arturo is correct? Estimate, then calculate.

If we round 185 to 200 and divide it by 5, we know in advance that the answer will be about 40 pages each day. This does not match Arturo's estimate.

Calculate to find the real answer.

$185 \div 5 = 37$ pages each day, which is close to our estimate.

Arturo needs to read 37 pages every day —not 20—or he will not finish in time.

A Choose the best answer.

Movie	Attendance
1st	227
2nd	198
3rd	176
4th	315

1. About how many people went to the first 2 shows?

Ⓐ 600   Ⓑ 500   Ⓒ 300   Ⓓ 400

Use estimation to find the combined attendance of the first 2 shows. 227 is rounded to 200 and 198 is also rounded to 200. 200 + 200 = 400. Choices A, B, and C are not reasonable estimates. Choice D is the correct answer. Use rounding to find the next answer.

2. About how many fewer people went to the 3rd show than the 4th show?

Ⓐ 100   Ⓑ 300   Ⓒ 200   Ⓓ 400

Answer the question. Show your work.

1. Oscar Chaplin III is one of the top weightlifters in the world. He recently won the United States championship. In the first event, "the snatch," he lifted 357 pounds; and in the "clean and jerk," he lifted 407 pounds. About how much did Chaplin lift in the two events?

2. Julia told her friend Alicia, that she estimated the Snake River to be about 2,000 miles long. Alicia said, "Actually, it's exactly 1,040 miles long." Explain why Julia's statement was not a good estimate.

3. There are about 753 students in Midland School. If the school includes only grades 2 through 6, about how many students are in each grade?

 Estimate the answer, then solve. Show your work on another sheet of paper.

1. James wanted to buy snacks for his party. Chips cost $5.75, soda costs $13.25, pretzels are $6.50, and cookies are $8.45. He estimates he will have enough money if he brings $30. Does he have enough money? If he does not, how much more does he need?

2. There were 1,692 fans at the first basketball game of the season. The third game attracted 1,278 fans. If about 5,000 fans attended the first 3 games, what is the least number of fans who attended the second game? What was the greatest number of fans who may have been there?

3. In the Atlantic Ocean, off the south shore of Long Island, New York, old ships are sunk to form manmade reefs. This is expensive, since each sinking costs about $5,000. These reefs attract small fish, which in turn attract large fish like tuna and sharks. One of the largest reefs covers 744 acres. Some are as small as 3 acres. If the average size of the reefs is 538 acres, how many acres do the 9 reefs cover?

L E S S O N   5

CHECK FOR REASONABLE SOLUTIONS

Estimation can help you check your answers to make sure they are reasonable and that they make sense.

In Buffalo, it rained 3.6 inches in March, 4.1 inches in April, and 5.4 inches in May. Kyle said that it rained about 35 inches that spring. Is his guess reasonable?

To estimate the solution, round 3.6 to 4, 4.1 to 4, and 5.4 to 5. Add the numbers to get about 13 inches. 35 inches is not a reasonable estimate.

Logic can also help you check how reasonable your solution is.

Eva wanted to buy a cookie for each of her 30 classmates. Her favorite cookies came in packages of a dozen. Eva asked her mom to buy 2 1/2 packages of cookies.

Eva's mom would tell her that her request was not reasonable. Why? Because stores do not sell 1/2 packages! Eva's mom would have to buy 3 packages of cookies.

 Choose the most reasonable answer.

1. If it takes Sal 5 days to read a book, how long will it take him to read 8 books?

Ⓐ 4 days   Ⓑ 40 days   Ⓒ 0.4 days   Ⓓ 400 days

Choices A and C are not reasonable answers, since both choices are less than the time it takes him to read one book. Choice D is far too great. The correct choice is B, which is found by multiplying 5 x 8 = 40. Do the next one on your own.

2. What is the perimeter of a rectangle with a width of 5 feet and a length of 9 feet?

Ⓐ 50 feet   Ⓑ 14 feet   Ⓒ 220 feet   Ⓓ 45 days

Answer the question. Show your work on another piece of paper.

1. Jake delivered 300 newspapers in 6 hours. He told his mother he delivered 5 newspapers per hour. Does Jake's answer seem reasonable? Explain your answer.

2. Esperanza measured the length of her bed. It measured 215 centimeters. She knew she needed to buy a cover at least 20 meters long. Is this a reasonable statement? Explain your answer.

3. Evan wrote a report about Greenland, an island about four times the size of France. Since it is mostly covered in ice, scientists are taking samples of the ice to learn about the Earth's climate in the past. In his report, Evan wrote that if Greenland is about 840,000 square miles, France is about 200,000 square miles. Is Evan's report accurate? Explain your answer.

4. Yamal bought cans of coffee for his office. He determined that if he bought a box containing 8 cans of coffee, and each weighed 13 ounces, the box would weigh about 7 pounds. His friend Kendra disagreed and said the box would weigh only 1 pound. Whose answer is more reasonable?

5. Claire and her family drove to Maine for their vacation. On the first day, they drove 382 miles and on the second day, they drove 293 miles. On the way home, they drove 105 miles further to do more sightseeing. Claire calculated they drove a total of about 1400 miles. Her brother, Nathan, said she was wrong, they had driven about 4,000 miles. Who had the more reasonable answer?

6. At the grand opening of Sayv-A-Lott store, every 10th customer got a discount coupon. In the first hour, there were 160 customers; and in the second hour, there are 120 customers. A total of 530 customers went to the store in the first 3 hours. Jan said that 290 customers received discount coupons in the third hour. Why is her statement unreasonable?

7. In four nights of reading last week, Umberto read 68, 65, 74, and 59 pages. Umberto estimated that he read about 400 pages in all. The actual number of pages is 266. How could he make a more accurate estimate? What would that estimate be?

SHARE YOUR THINKING

Sometimes you may be asked to explain how you solved a math problem. But how do you use words to explain numbers? Look at the following examples.

Richard wanted to stack boxes so that the lightest box was on top and the heaviest was on the bottom. The boxes weighed 3.4 pounds, 3.04 pounds, 4.3 pounds, and 0.43 pounds. In what order will he stack the boxes from top to bottom?

Traci's explanation:

1) Richard should look first at the one's place in each number. He should group numbers with the same ones digit together. Then he should put those numbers in order.

2) Richard should compare the numbers with the same ones digit by looking at tenths.

3) So the numbers in order from least to greatest (or top to bottom) are: 0.43, 3.04, 3.4, 4.3

The Diazes ordered a pizza for dinner. Consuelo ate 1/4 of the pizza, Marco ate 1/3. How much of the pizza did they eat?

Gavin's explanation:

1) Find the lowest common multiple of the denominators (4, 3; LCM = 12)

2) Rename each fraction using the common denominator (3/12, 4/12)

3) Add the numerators together, write the answer above the common denominator (3 + 4 = 7, so 3/12 + 4/12 = 7/12)

4) Check to see if the answer can be simplified. (It can't.)

5) So, the answer is "Consuelo and Marco together ate 7/12 of the pizza."

A Which explanation best fits the question?

1. Jennifer is sorting her sticker collection into 6 boxes. She has 118 stickers. If she puts the same numbers in each box, about how many stickers will be in each box?

Ⓐ Round 118 to 120. Divide 120 by 6, which is 20 stickers.

Ⓑ Round 118 to 100. Divide 100 by 6, which is 16 stickers with 4 left over.

Ⓒ Round 118 to 120. Divide 120 by 6, which is 18 stickers.

Ⓓ Divide 118 by 6, which is 19 stickers with 4 left over.

Choice A correct. The question calls for a close estimate. Try the next on your own.

B Is the explanation of the solution to the problem correct? Write a sentence or two to explain your opinion.

1. Buck has $5.00. He wants to buy some muffins and donuts for his family. A muffin costs $.75 and a donut costs $.50. Buck wants to buy 2 muffins and spend the rest of his money on donuts. How many donuts can he buy?

   Explanation: To find the number of donuts he can buy, Buck should multiply the cost of a muffin by 2 and subtract that cost from $5.00. The remaining amount should be divided by the cost of a donut.

2. Mercury is the planet closest to the sun. One day on Mercury lasts as long as 176 days on Earth. How long does it take to complete one day/night cycle on Mercury?

   Explanation: To find the number of hours in a day/night cycle on Mercury, divide 176 by 24.

C Solve the problem. Explain your steps on another sheet of paper.

1. In a recent survey, one-fourth of all the students in the fifth and sixth grades chose blue as their favorite color. If 80 students chose blue, how many students are in the fifth and sixth grades?

2. Have you ever heard of Ruth Handler? In 1938, she got married, and 4 years later Ruth teamed with her husband to manufacture picture frames. They soon started making toys, and 17 years later, she created the Barbie Doll. Today, the doll is sold in 150 countries. In what year was the Barbie doll created?

SKILL TUNE-UP: MULTIPLICATION AND DIVISION

Multiply by 1-digit numbers

First multiply the ones, and regroup if needed. Then multiply the tens, add any number carried over, then regroup if needed. Finally multiply by hundreds, add any number carried over.

	3	13	13	
239	239	239	239	239
x 4	x 4	x 4	x 4	x 4
	6	56	1,356	1,356

Multiply by 2-digit numbers

First multiply by the ones digit. Next multiply by the tens digit. Finally add the products together

	2	1	49
49	49	49	x 23
x 23	x 23	x 23	147
	147	147	+ 980
		980	1,127

Divide

Multiply, then subtract. Bring down the ones. Divide the ones, multiply, then subtract. If needed, write the remainder. Check to be sure the remainder is smaller than the divisor.

	1	1	11	11 R6		2	2	23R15
7)83	7)83	7)83	7)83	7)83	23)544	23)544	23)544	23)544
	- 7	- 7	- 7			- 46	- 46	- 46
	1	13	13			8	84	84
			- 7				- 69	- 69
			6				15	15

A **Choose the correct answer.**

1. 78 × 67 =

   Ⓐ 4,776   Ⓑ 5,226   Ⓒ 4,826   Ⓓ 6,466

   The correct choice is B. Try the next problem.

2. 831 ÷ 17 =

   Ⓐ 51   Ⓑ 40 R15   Ⓒ 48   Ⓓ 48 R15

B **Write a sentence or two to explain your answer.**

1. Ruth's favorite baseball team, the San Francisco Giants, scores an average of 4 runs per baseball game. She predicts that in a 162-game season, they will score a little over 40 runs. Does her answer make sense?

2. There are 68 students trying out for the school play. They will divide into groups of 7 for the audition. The remaining 5 students will form their own group. Does this seem like a reasonable solution such that all students will be part of a group?

C **Solve. Show your work on another piece of paper.**

The Eiffel Tower in Paris, France, was built in 1889 on the 100th anniversary of the French Revolution. Soaring to a height of 1,052 feet, it was considered a miracle of construction with 18,038 steel pieces. To walk to the top, you will have to climb 1,665 steps. Today, it is one of the most popular tourist spots in the world with 5,530,279 visitors in 1996.

1. In buildings, each story is 12 feet in height. About how many stories is the Eiffel Tower?

2. Every 7 years, 50 tons of dark brown paint are used to protect the structure. How many tons of paint will be used in 98 years?

STRATEGY: SOLVE A SIMPLER PROBLEM

Solving a simpler problem first can sometimes help you solve a math problem.

When multiplying numbers with zeroes, first multiply the non-zero digits, then add the number of zeros to the product.

40 x 30 = 1200   80 x 50 = 4000

When dividing numbers with zeroes, you can cross out an equal number of zeroes in the dividend and the divisor.

15̸0̸ ÷ 3̸0̸ = 5   2,70̸0̸ ÷ 9̸0̸ = 30

Round the numbers to make the problem simpler. Solve, then follow the same steps to solve the original problem. This also gives you an estimate to compare with your actual answer.

The track team runs around a square block with each side measuring 9.5 meters. To reach their goal of 150 meters, how many times will the team need to run around the block?

150 ÷ 10 = 15 times(approximately)
150 ÷ 9.5 = 15.8 times

 Which is the best way to simplify the problem?

1. A factory produces an average of 20,000 CDs per day. How many CDs do they produce in 60 days?

   Ⓐ multiply 2 x 6 and add 4 zeroes

   Ⓑ divide 6 by 2 and add 5 zeroes.

   Ⓒ multiply 2 x 6 and add 5 zeroes

   Ⓓ multiply 20,000 x 60 and add 5 zeroes

To simplify, eliminate all the zeroes and multiply. 5 zeroes are eliminated, so choice A is incorrect. The correct operation is multiplication, not division, so B is incorrect. No zeroes have been eliminated in choice D, so it is incorrect. Choice C is the correct way to simplify the problem. Do the next one on your own.

2. Books are on sale for $4.95 each and magazines for $2.25. Frank has $15 to spend. If he wants to buy 2 books and 3 magazines, does Frank have enough money?

    Ⓐ Round $4.95 to $5 and multiply by 3. Round $2.25 to $2 and multiply by 2. Add the two products together.

    Ⓑ Round $4.95 to $5 and multiply by 2. Round $2.25 to $3 and multiply by 3. Add the two products together.

    Ⓒ Round $4.95 to $4 and multiply by 2. Round $2.25 to $2 and multiply by 3. Add the two products together.

    Ⓓ Round $4.95 to $5 and multiply by 2. Round $2.25 to $2 and multiply by 3. Add the two products together.

B   Simplify to solve. Show your work on another piece of paper.

1. Flights from Boston to San Diego, California are $354 round trip. How much will it cost for 3 people to fly round trip from Boston to San Diego?

2. Carla collected 590 cans for the recycling drive at her school. If she took 10 days to collect the cans, how many cans did Carla collect each day?

3. Exercise is a great way to stay healthy. It keeps the heart and lungs working their best. Studies have shown that running for an hour, a person can expect to use 720 calories. Other studies have shown that walking for the same length of time, you can expect to burn 324 calories. Last week, Terri walked for 4 hours. If her husband, Tom ran for 2 hours, who used up more calories?

4. Sara wants to try out a new Tofu and Vegetable Stir Fry recipe for dinner. To serve 4 people, she needs two 10.3 ounce bags of mixed stir fry vegetables along with 2 packages of extra-firm tofu. If she expects to serve dinner for 8, how many ounces of stir fry vegetables will Sara need to buy?

5. Daryl wanted to put some decorative rocks in his garden. He bought a 60-pound bag. As he placed the rocks in his garden, he counted 18 rocks. How much did each rock weigh?

STRATEGY: WRITE AN EQUATION

Writing an equation first can help you solve a story problem. Writing equations can be especially helpful if a problem has many numbers or many steps.

Ms. Woods actually buys boots for her pets. If Ms. Woods has 4 dogs, 5 cats, and 2 pot-bellied pigs, how many boots will she buy this season?

Let b stand for the number of boots. Then, $b =$

   (4 dogs x 4 feet)
+ (5 cats x 4 feet)
+ (2 pigs x 4 feet)

Here are two ways to write the equation. Both result in the correct answer.

$4 (4 + 5 + 2) = b$

$4 (11) = b$

$44 = b$

$(4 \times 4) + (5 \times 4) + (2 \times 4) = b$

$(16) + (20) + (8) = b$

$44 = b$

Ms. Woods will buy 44 boots for her pets.

A   **Which equation best represents the problem?**

1. Melanie bought 3 pies for $5 each. If she paid with a $20 bill, how much change did she get back?

  Ⓐ $20 - (3 \times 5) = ?$ Ⓒ $(20 \times 3) - 5 = ?$

  Ⓑ $(20 - 3) \times 5 = ?$ Ⓓ $(5 \times 3) - 20 = ?$

The correct choice is A since $20 - $15 = $5 in change. Find the next equation on your own.

2. Trevor rode his bike 8 miles each day for 6 days. Keegan rode his bike 7 miles a day for 9 days. How many more miles did Keegan ride than Trevor?

  Ⓐ $(8 \times 6) - (9 \times 7) = ?$ Ⓒ $(9 + 7) - (8 + 6) = ?$

  Ⓑ $(9 \times 7) - (8 \times 6) = ?$ Ⓓ $(9 \times 7) + (8 \times 6) = ?$

B **Write an equation that you could use to solve the problem.**

1. After the 2001 – 2002 NBA season was completed, Clifford Robinson of the Detroit Pistons received 5 first place votes. Kobe Bryant of the Los Angeles Lakers received 7 more first place votes than Robinson. If Kevin Garnett of the Minnesota Timberwolves received 2 fewer first place votes than Robinson, how many first place votes did each player get?

2. Two people worked on a construction project. The first worker was paid $15 an hour for working 6 hours. The other worker was paid $17 an hour and worked 8 hours. What is the total amount the two workers were paid?

3. Granite is used in the construction of many buildings. It resists weathering for a long time and still looks beautiful. One construction company had 20,000 pounds of granite left over from their work. This was 200 pounds more than half the amount they started with. How much granite did they start with?

 **C** **Write an equation, then solve. Show your work on another piece of paper.**

1. A new process is being used to filter the blood of patients getting kidney transplants. This was important for the 38,381 patients awaiting transplants in 2000. In the past, many patients died after the operation because there were problems with their blood. If 13,372 patients received kidney transplants in 2000, how many people were still awaiting new kidneys at the beginning of the year 2001?

2. In the first 4 months of this year, it rained 12 inches. The normal amount of rainfall for the first 4 months is 20 inches. How much less than the monthly average did it rain so far this year?

3. Ellie studied for 2 hours a night on Monday and Wednesday last week. She also studied 4 hours on Saturday. If her goal was to study 10 hours last week, did she meet her goal? If not, how many more hours does Ellie need to study?

4. The temperature was 78°F at 4:00 P.M. It had risen 2° an hour for the 3 previous hours. What was the temperature at 1:00 P.M.?

STRATEGY: USE INVERSE OPERATIONS

Some problems can be solved more easily if we keep in mind that multiplication and division are inverse operations. Addition and subtraction are also inverse operations.

Rafiq wanted to reduce the dimensions of a scanned photo in half to fit a webpage. If his new picture was 4 inches by 5 inches, how big was the original photo?

What we know The original photo size was divided by 2. The resulting photo wa 4 by 5 inches.

Our plan Use inverse operations. Multiply the dimensions of the resulting photo by 2.

Calculate Let p = photo size.

p = 2 (4 inches by 5 inches)

p = 8 inches by 10 inches

Check Now, think about the problem in reverse: A photo, sized 8 inches by 10 inches, reduced by 2, is 4 inches by 5 inches. The original photo was 8 inches by 10 inches.

 A   Choose the correct answer.

1. Jefe is thinking of a secret number. If you subtract 2, add 5, and then subtract 7 the resulting number is 16. What is his secret number?

   Ⓐ 16   Ⓒ 18

   Ⓑ 22   Ⓓ 20

   Choices A, B, and C are incorrect. To solve this problem, use the resulting number (16) along with opposite operations, as follows: $16 + 7 - 5 + 2 = 20$. The correct choice is D. Try the next problem on your own.

2. Dan's teacher told the class to read a certain number of books between October and April. Dan read 3 books each month in February, March, and April. He read 5 books in January, and 2 books each month in November and December. If he still needs to read 7 books to reach his goal, how many books is Dan expected to read?

   Ⓐ 18   Ⓒ 25

   Ⓑ 30   Ⓓ 11

B Write an equation to use in solving the problem.

1. Janet started working at the mall after school to save money for new clothes. She made $65 the first week and $80 the second week. If she now has $400 saved, how much did Janet have before she started working?

2. Many people suggested ways to reduce the cost and improve the safety of the Space Shuttle. Ideas such as having the shuttle take off like a plane may reduce the cost of projects such as delivering objects to the Space Station to $1,000 per pound. If this is one-tenth the current cost, how much does it now cost to deliver equipment?

3. Pete went shopping for groceries. He bought a loaf of bread for $2.29, a container of orange juice for $1.95, a box of cereal for $3.50, and a package of cookies for $1.75. If he was given $3.25 in change, how much money did he give the cashier?

C Solve. Show your work on another piece of paper.

1. It takes the planet Uranus about 84 years to complete one orbit around the sun. If you think this is long, you'll be surprised to learn that Uranus' orbit is only half as long as the orbit of Neptune, and one-third the orbit of Pluto. About how long does it take Neptune and Pluto to orbit the sun?

2. Mary now has 80 CDs. Last year she received one-quarter of these for her birthday. She bought another 15 CD's last month. How many CD's did Mary have before her birthday?

3. After a day of sightseeing, Tony returned to his hotel at 6:00 P.M. If he spent 3 hours in museums, 2 hours eating meals, and 4 hours shopping, what time did Tony leave his hotel in the morning?

4. The Barr family traveled 1,537 miles on their vacation. They first drove from their home in Wichita, Kansas to St. Louis, Missouri. They then rode 312 miles to Nashville, Tennessee. From Nashville, the Barrs drove 780 miles back home. How far is it from Wichita to St. Louis?

CHOOSE A STRATEGY

Review the strategies you have learned so far.

Solve a Simpler Problem

Multiply the non-zero digits, then add the number of zeros to the product.

To construct a patio, Roberta's family used 500 stones, weighing 20 pounds each. How many pounds of stone had they bought?

500 x 20 = 10,000 pounds

Round the numbers to make the problem simpler. Solve the new problem, then follow the same steps to solve the original problem.

If your heart beats 87 times per minute, how many times will it beat in an hour?

equation: 87 x 60 = b
simplify: 90 x 60 = 5,400
actual answer: b = 5,220

Write an Equation

Writing an equation can help decide the steps to solve.

On Mars, Ryan weighs 30 pounds and Jackson weighs 28 pounds. On Earth, they would weigh 3 times more. How much do they weigh altogether on Earth?

3 (30 + 28) = w or (3 x 30) + (3 x 28) = w
w = 174 pounds

Use Inverse Operations

Remember that multiplication/division and addition/subtraction are inverse operations.

Mr. Ramone spent $24 on bags of popcorn for each of his 6 children. How much did each bag cost?

p x 6 children = $24

p = 24 ÷ 6

bags of popcorn cost $4 each

A Which strategy would be most helpful in solving the problem?

1. Carl read an average of 76 pages each night for 12 nights. About how many pages did Carl read?

   Ⓐ work backwards   Ⓑ write an equation   Ⓒ cross out zeroes   Ⓓ estimate

2. Alice gave 15 pieces of candy to James and some to Felicia. If she started with 50 and has 12 left, how many pieces of candy did Alice give to Felicia ?

   Ⓐ work backwards   Ⓑ write an equation   Ⓒ cross out zeroes   Ⓓ estimate

B Describe a good strategy to solve the problem. Explain why you chose the strategy you did.

Earth's Layers

Crust	65 kilometers
Mantle	2,250 kilometers
Outer Core	2,900 kilometers
Inner Core	?

1. If the combined thickness of the Earth's four layers is 6,421 kilometers, how thick is the inner core?

2. What is the difference in thickness between the mantle and outer core?

3. Scientists found that the type of rocks in the mantle changed every 50 kilometers. How many different rock types did they probably find?

C Solve. Show your work on another piece of paper.

1. On May 2, 2002, Mike Cameron of the Seattle Mariners became the 13th Major League baseball player to hit 4 home runs in a game. The first player to accomplish this difficult feat was Robert Lowe in 1894. Another player hit four home runs two years later, but it was another 36 years before Lou Gehrig did this again. When did Gehrig hit four home runs in a game?

2. A large amount of water is used every day. Of course you're aware that we use water to bathe and drink. You may not be aware that for every car you see, 1,000 gallons of water was used to make it. How many gallons of water would be used to make 500 cars?

3. Parking cost $4. There were 23 cars were in the parking lot. However, 10 people had parked for free because they were helpers. How much money did the parking lot attendant collect?

MULTI-STEP STORY PROBLEMS

 A Choose the correct answer.

1. Jack earned $66 at his after-school job. His mother gave him $25 for his birthday. If he spent $45 on clothes, about how much did Jack have left?

   Ⓐ $80   Ⓑ $70   Ⓒ $60   Ⓓ $50

2. Laquinta bought 3 bags of apples. Each bag weighed 5 pounds. She also bought 2 bags of rice weighing 3 pounds each. What was the total weight of Laquinta's purchase?

   Ⓐ 21 pounds   Ⓑ 13 pounds   Ⓒ 25 pounds   Ⓓ 20 pounds

3. Skyway Travel Agency is offering a special cruise package for people paying two months in advance. The cost per person is $800 for a five-day cruise. All 10 members of the Lopez family are planning to go. Which would be the best strategy to use to figure out the total cost for the Lopez family?

   Ⓐ use inverse operations   Ⓒ cross out zeroes

   Ⓑ estimate   Ⓓ write an equation

4. Together, Trena and her sister Alicia weigh 143 pounds. If Trena weighs 68 pounds, how much does Alicia weigh?

   Ⓐ 75 pounds   Ⓑ 65 pounds   Ⓒ 70 pounds   Ⓓ 80 pounds

B Write a sentence or two to answer the question.

1. In a mixed bag of 300 nuts, about 93 nuts are cashews and 47 nuts are pecans. Arnaldo said the best strategy to find the number of nuts that are neither cashews nor pecans is to work backwards. Is Arnaldo correct? Explain your answer.

2. In Pennsylvania, the greatest distance from its eastern to western borders is 306 miles. At its widest point, Pennsylvania is 175 miles from north to south. If area is found by multiplying length times width, how would you estimate the area of Pennsylvania?

C Solve. Show your work and explain your strategy on another piece of paper.

We use oil for many things—to heat homes and to make gasoline for cars, trucks, jets, and other vehicles. Did you ever think of how we get oil? Oil is found by drilling deep underground. It is also found offshore under the ocean. Large platforms, known as rigs, are constructed so this natural resource may be brought to the surface.

1. A small submarine may carry workers deep underwater to make repairs to the drilling pipe. If the submarine descends 1,892 feet and there is still 1,213 feet between the submarine and the ocean floor, how deep is the water?

2. In 1890, before the automobile was invented, oil production was at a low level. By 1920, production had increased by 612,000,000,000 barrels to 689,000,000,000 barrels of oil. By 1930, as the popularity of cars soared, the total had risen by 723,000,000,000 barrels a year. What was the yearly production of barrels of oil in 1890 and 1930?

3. When they drilled for oil, companies also discovered natural gas. Before they realized its value, great amounts were burned and wasted. In the 1920s, huge pipelines were constructed to carry the natural gas great distances from the oil fields to cities and towns. Usually, 20 miles of pipeline can be built in 2 weeks. If a pipeline connects to a city 1,081 miles away, about how many weeks will it take to construct the pipeline?

4. In 1880, 30,000,000,000 barrels of oil were produced. The United States supplied 80% (0.8) of the world's oil. How many barrels of oil did the USA produce in 1880?

5. In its best year, Eastland County, Texas, produced 22,000,000 barrels of oil. Stephens County, Texas, produced 1,400,000 barrels of oil in its best year. Which county was more productive in its best year? By how much?

SKILL TUNE-UP: WORKING WITH DECIMALS

A decimal is a number with at least one digit to the right of the decimal point. Like a fraction, a decimal number shows part of a whole.

A number smaller than 1 can be written as a decimal. Sometimes a zero holds the ones place, but not always.
0.7 = .7

Add, subtract, multiply, and divide numbers with decimals the same as you would numbers without decimals. But remember to keep the decimal aligned and in the correct place.

```
  .5
 1.3
+1.07
 2.87
```

```
        7  9  10
$8. 0̶ 0̶
−$0.9 9
$7.0 1
```

```
  8.3     1 decimal place
x 0.2     1 decimal place
 1.66     2 decimal places
```

```
      2.4
.6⟌1.44     Move the
  -1 2       decimal the
    24       same number
  − 24       of places to
     0       the right in
             the divisor and
             dividend
```

 A Choose the correct answer.

1. **3.72 + .9 =**

   Ⓐ 3.71   Ⓑ 3.81   Ⓒ 3.62   Ⓓ 4.62

Compare your work with the solution below to see that D is the correct answer. Try the rest on your own, making sure to align decimals.

```
   1
 3.72
+.90
 4.62
```

2. **$9.00 − $0.68 =**

   Ⓐ $8.32   Ⓑ $8.99   Ⓒ $9.42   Ⓓ $7.95

3. 5.5 x 2.6 =

Ⓒ 143   Ⓓ 14.3   Ⓔ 1.43   Ⓕ .143

4. 14.8 ÷ 3.7

Ⓒ .4   Ⓓ 4.4   Ⓔ 40   Ⓕ 4

5. 2.56 ÷ 1.6

Ⓒ .16   Ⓓ 0.16   Ⓔ 1.6   Ⓕ 160

 B   **Solve. Show your work on another piece of paper.**

1. Janine pays $19.95 per month for her Internet service provider. She told her mother she plans to pay $1,197.00 for six months. Show why Janine will not need to pay so much.

2. Immigrants to the United States come from many places. The table below shows how many immigrants came to the United States between 1820 and 1992.

Region	Millions of Immigrants
Asia	6.8
Central and South America	7.5
Europe	37
Africa	0.4

Ken said that the total immigration to the United States from all these regions was 51.7 million. Was Ken correct? Explain your answer.

3. Carlisle had $10 to spend on a birthday present for his friend Bijan. He bought a card for $2.95 and some candy for $5.75. How much change did Carlisle get back?

4. The albatross is a large bird living near the ocean. For years, nobody knew where the birds went when they left for long trips over the ocean. Recently, scientists attached tiny transmitters to some birds to find exactly where they were going. The results showed most of their trips were to find food, and their flying speeds ranged up to 85 miles per hour. If an albatross traveled 64.5 miles in 1.5 hours, what was its average speed?

STRATEGY: ESTIMATING

Sometimes dividing a whole number results in a quotient that is a decimal. Here is an example:

In one hour, four friends earned $26 washing cars. How much did each person receive?

```
      $6.50
4/$26.00
  − 24
    20
  − 20
     0
```

Sometimes you need to round a decimal number. Read the question carefully to decide how to round your answer.

Round to the nearest whole number.

Mr. Rossi needed fabric for his 27 students to create their own flags. He wants the flags to be about 0.3 yards long. What length of fabric did he buy?

```
   27
x .3
  8.1
```

Would anyone notice – or care – if the flags aren't exactly the same size? Mr. Rossi probably bought 8 yards of cloth instead of 8.1 yards.

Round to the nearest whole number.

Delia needed to get money from the ATM to buy presents. How much does she need to take from her account? (Remember, ATM machines give out money in multiples of $10.)

Gift List	
Mom – silk scarf	$ 9.95
Dad – new software	$14.50
Grandma – lunch	$10.00
Jacob – science kit	+ $ 7.99
	$42.44

Delia will need to take out at least $50.

Round to the nearest hundredth.

The 4 car washers earned $86.50 altogether one day. How much did each person receive?

```
   $21.62 R 2
4/$86.50
 − 8
   06
 − 4
   25
 − 24
   10
 −  8
    2
```

For some problems, you might continue dividing. But you already know that money amounts only use 2 decimal places.

The friends will just have to agree to let someone have the extra 2 cents.

Choose the best answer.

1. The sixth-grade teachers wanted to have an adult chaperone for each group of 6 students going on the field trip. Altogether, 39 sixth graders went on the field trip. How many adults went along as chaperones?

   Ⓐ 6   Ⓑ 6.0   Ⓒ 7   Ⓓ 6.7

2. A pound of cherries costs $.41. If Adriana buys 0.7 pounds, she will need at least—

   Ⓐ $.32   Ⓑ $.30   Ⓒ $.28   Ⓓ $.29

B **Solve. Show your work.**

1. Ellen had 5 books that weighed 22 pounds. What was the average weight of a book?

2. What number, using the digits 3, 1, and 5, can be rounded to 14 ?

3. What number, using the digits 1, 3, and 5, can be rounded to 2?

4. Louisa is writing a report about trees. She summarized part of her research in the chart below.

Tree	Diameter (Meters)
Hickory	0.9
American Elm	1
Quaking Aspen	0.8
Bald Cypress	1.5

What is the average diameter of these four trees?

STRATEGY: SOLVE A SIMPLER PROBLEM

Solving a simpler problem first can sometimes help you solve a story problem.

The Gursky family needs to buy crayons for each of their seven children. The 4 little Gurskys need small packs, each costing $2.99; the 3 big Gurskys need large packs, which cost $4.65 apiece. How much will the Gurskys pay for crayons altogether?

To solve a simpler problem, round the decimals to the nearest whole number. Solve the new problem, and then follow the same steps to solve the original problem if an exact answer is needed.

$2.99 rounds to $3, and $4.65 rounds to $5. Then:

$(4 \times \$3) + (3 \times \$5) = t$

$t = \$27$

Now you can use the actual numbers in your equation.

$(4 \times \$2.99) + (3 \times \$4.65) = t$

$t = \$25.91$

Finally, compare your estimated answer with the actual answer to check yourself.

$27 is close to $25.91

A **Which is the best way to simplify the problem?**

1. Raphael's grandfather offered to pay him $4.40 an hour for painting their garage. Raphael spent $2\frac{1}{2}$ (2.5) hours painting on Saturday. About how much did Raphael earn on Saturday?

   Ⓐ $3 \times 5 = a$ Ⓑ $2 \times 4 = a$ Ⓒ $3 \times 4 = a$ Ⓓ $2 \times 5 = a$

   Choice C is the correct choice, since 2.5 rounds to 3 and 4.4 rounds to 4. Now, do the next problem on your own.

2. Glenn Intermediate School has 396 students. If each lunch table holds an average of 4.8 students, how many lunch tables are there?

   Ⓐ $400 \div 5 = a$ Ⓑ $300 \div 5 = a$ Ⓒ $400 \div 4 = a$ Ⓓ $300 \div 4 = a$

B Read the story problem. Write a sentence or two telling how to solve it using the strategy "solve a simpler problem."

1. Carrie saved $27.25 of her allowance last month and $23.45 this month. Her friend, Jenny, saved $34.80 and $16.50. Who saved more money?

2. Theodore Roosevelt, the 26th President, had a summer home from 1901 to 1909 on Oyster Bay, Long Island. He held huge parties and had one with 8,000 guests. If each of the guests ate an average of 2.7 cookies and 1.3 cups of lemonade, how many cookies and cups of lemonade were served?

3. Emma is building a fence around her yard. The yard is 40.8 feet wide and 81.6 feet wide. If she puts a post every 5.1 feet, how many posts will Emma need?

C Solve. Show your work on another sheet of paper.

1. In 1948, Earl Shaffer accomplished an amazing feat. He was the first person to walk the entire length of the Appalachian Tail, which stretches 2,058 miles from Georgia to Maine. In 1965, he hiked from Maine to Georgia, and became the first person to hike the trail in both directions. In 1998, at the age of 79, Earl hiked the length of the trail once again. If Earl averaged 16.5 miles per day, how many days did it take him to hike one way?

2. On May 18, 1980, Mount St. Helens, a volcanic mountain in the state of Washington, erupted with a great explosion. When the eruption first occurred, huge amounts of ash and pumice (volcanic rock) covered much of the mountain and the surrounding area. At first, the pumice was over 1,003.6 degrees Fahrenheit. Fourteen days later, some of the pumice was as still as hot as 779.8 degrees. How many degrees per day did the pumice cool?

3. Patrick bought supplies for his family's camping trip. He bought 3 sleeping bags at $24.95 each, 2 lanterns costing $15.25 apiece, and a tent for $168.75. If he had $250, did Patrick have enough money to pay for the camping supplies? If not, how much more did he need?

STRATEGY: LOOK FOR A PATTERN

1 ● Gather Information
2 ● Plan
3 ● Calculate
4 ● Check

Learn to recognize repeating decimals.

$3 \div 9 = ?$

$$\begin{array}{r} .333333333 \\ 9\overline{)3.0} \\ -2\,7 \\ \hline 3.0 \\ -2\,7 \\ \hline 3.0 \\ -2\,7 \\ \hline 3 \end{array}$$

The quotient will repeat to infinity. When you find a repeating pattern, simply write a bar over the repeating number. Just make sure that you have done enough of the problem to know when the number repeats.

$3 \div 9 = 0.\overline{3}$

$7 \div 6 = ?$

$$\begin{array}{r} 1.16666666 \\ 6\overline{)7.0} \\ -6 \\ \hline 10 \\ -6 \\ \hline 40 \\ -36 \\ \hline 40 \\ -36 \\ \hline 40 \\ -36 \end{array}$$

$7 \div 6 = 1.1\overline{6}$

Notice patterns that can help you find a "shortcut" to a solution. For example:

$4 \div 10 = 0.4$ $0.05 \div 10 = 0.005$
$80 \div 10 = 8.0$

When dividing by ten, move the decimal point in the dividend one place to the left to find the quotient.

A Choose the correct answer.

1. Mateo has a string that is 5 yards long. If he cuts it into 3 equal pieces, how long will each piece be?

   Ⓐ $1.\overline{06}$ yards   Ⓑ $1.\overline{6}$ yards   Ⓒ $0.\overline{6}$ yard   Ⓓ $1.\overline{7}$ yards

Compare your work to the problem below to see that B is correct. Do the next one on your own.

```
        1.666
   3) 5
    - 3
      20
    - 18
      20
    - 18
      20
    - 18
       2
```

2. Mr. Franklin gave his children $7 so each child could get a bottle of juice. He has 10 children. How much can each child spend on juice?

Ⓐ $70　　　Ⓑ $.07　　　Ⓒ $0.65　　　Ⓓ $.70

B　**Write a phrase or sentence to answer the question.**

1. There was a heavy rainstorm in Chicago yesterday. 3.2 inches of rain fell in 10 hours. How would you find the average hourly rainfall? What shortcut could you use to help you find a solution?

2. Sylvia gives homemade chocolates as gifts. She made an 8-pound batch of chocolate to divide equally among 6 boxes. Sylvia calculates that each box will contain 0.75 pounds of chocolate. Why is Sylvia wrong? What is the correct answer?

3. Explain how you could use a shortcut when dividing a decimal by 100 and by 1,000.

CHOOSE A STRATEGY

Review the strategies you have learned so far. Remember, there is more than one way to solve some math problems!

Estimating—If you round a decimal, decide the place to round up or down.

Mrs. Hassan gave her 8 children $35 for lunch. How much could each child spend?

Each child could have $4.37, with 4 cents left.

$$
\begin{array}{r}
\$4.375 \\
8\,\overline{)\$35.00} \\
-32 \\
\hline
30 \\
-24 \\
\hline
60 \\
-56 \\
\hline
40 \\
-40 \\
\end{array}
$$

Solve a Simpler Problem—Round numbers to make a problem simpler. Solve the new problem, then follow the same steps to solve the original.

To make a castle model, Micah used 5.8-inch pieces of wood for the 8-sided tower, and 9.3-inch pieces for the 4 regular walls. What is the total length of the wood Micah used?

Round the numbers: 5.8 to 6 inches, 9.3 to 9 inches

Equation: (8 x 6 in) + (4 x 9 in) = 84 inches

Using actual numbers: (8 x 5.6) + (4 x 9.3) = (44.8) + (37.2) = 82 inches

Look for a Pattern – Watch for repeating decimals.

Six friends bought a cheesecake with 8 pieces. How much of the cake will each person eat?

Each will get 1 piece and about 0.3 (about 1/3) of another.

$$
\begin{array}{r}
1.\overline{3} \\
6\,\overline{)8} \\
-6 \\
\hline
20 \\
-18 \\
\hline
20 \\
-18 \\
\hline
2 \\
\end{array}
$$

Look for "shortcuts."

To organize the street sale, Sheng asked each family pay him a percentage of what they earned. He made this list after the sale was over.

What Each Family Owes Me
McIntosh - $35.00 x 0.1 = $3.50
Johnston - $24.50 x 0.1 = $2.45
Dion - $89.00 x 0.1 = $8.90
Cardiff - $76.50 x 0.1 = $7.65
Winarni - $115.50 x 0.1 = $11.55

A shortcut for multiplying by 0.1 could be to move the decimal one place to the left.

A Choose the correct answer.

1. If it rained 1.05 inches each day for 5 days, about how much did it rain?

    Ⓐ 7 inches   Ⓑ 4 inches   Ⓒ 6 inches   Ⓓ 5 inches

2. If 10 boxes of cookies cost $15.50, how much did each box cost?

    Ⓐ $15.50   Ⓑ $1.50   Ⓒ $1.55   Ⓓ $0.155

3. If a 10-inch piece of wood is divided into 12 pieces, each piece will be—

    Ⓐ 0.83 inches   Ⓑ 0.82 inches   Ⓒ 0.835 inches   Ⓓ 0.844 inches

 B Solve. Explain your answer.

1. During his first week of training, Carlo ran 12.71 miles. The next week, he increased his mileage to 13.09, and the third week, Carlo ran 13.65 miles. He estimated he ran 40 miles in 3 weeks. Was Carlo correct? Explain your answer.

Kendra wants to buy a new camera. She sees an ad from The Ace Camera Company for its new digital camera, the XL7, selling for $279.99. It is very fast, starting up in just 1.8 seconds. The 1.5-inch LCD monitor is one of the clearest and largest ever.

2. The XL7 is easy to carry, weighing much less than many other digital cameras. In fact, when 100 are shipped to the store, they weigh just 550 ounces, or 34 pounds, 6 ounces (excluding the weight of the box). What is the weight of each XL7?

3. Kendra is planning to buy a second XL7 camera as a gift. She will also buy the extra carrying case for each camera, for $13.45 apiece. If Kendra has budgeted $600 to buy the equipment, does she have enough money? Round the numbers and use them to explain your answer.

L E S S O N   6

MULTI-STEP STORY PROBLEMS

A Circle the letter of the correct answer.

1. Mei works hard to make his homemade bean soup. Some beans are boiled, others are stir fried, and others are both boiled and fried. To make the soup, Mei split 8 ounces of dried beans into 15 equal piles. How much does each pile weigh?

   Ⓐ 0.53 ounces   Ⓑ 5.3 ounces   Ⓒ 0.52 ounces   Ⓓ 0.54 ounces

2. On a recent trip from Boston to Virginia, the Smith family drove 537 miles in 10 hours. What was their average speed in miles per hour?

   Ⓐ 537   Ⓑ 5.37   Ⓒ .537   Ⓓ 53.7

3. The population of Michigan was 9.94 million in the 2000 census, while the population of Hawaii was 1.21 million. What was the difference in their populations?

   Ⓐ about 7 million   Ⓒ about 9 million

   Ⓑ about 10 million   Ⓓ about 8 million

4. One mile is equal to about 1.6 kilometers. If Crystal ran 16 kilometers, about how many miles did she run?

   Ⓐ 100 miles   Ⓒ 1 mile

   Ⓑ 10 miles   Ⓓ 1.01 miles

B Use the information below to solve.

With each increase in elevation, temperature decreases. The table shows the temperature (in degrees Fahrenheit) for each 2,000-foot gain in elevation (land height above sea level).

Elevation (in feet)	Temperature (° F)
Sea level (0)	78.5
2,000	71.5
4,000	64.5
6,000	57.5
8,000	50.5
10,000	43.5

1. Claire said that there is a 21-degree decrease in temperature between 2,000 and 6,000 feet. Is she correct? Explain your answer and use estimation as part of your explanation.

2. There is a 3.5-degree change in temperature for every 1,000-foot change in elevation. This works out to a 0.35-degree change for every 100 feet. Explain why this is true.

3. What would the temperature be at 1,000 ft? Explain how you found the answer.

 Use the information below to solve.

Devon wrote a report on minerals. He found that minerals are mined in many places around the world. For example, there are large copper mines in North America, Africa, and Central Asia. About 7.7 million tons of copper are mined each year. Other minerals, such as zinc and lead, are also produced in these areas. Each year, 6.2 million tons of zinc and 3.8 million tons of lead are mined and produced. Without these valuable natural resources, many products we use every day would not be available.

1. If the rate of production is steady, how many tons of zinc and lead are produced each month?

2. Other minerals, such as aluminum, are used in products such as containers, aircraft, and automobiles. About 18 million tons a year are mined. It is very important to recycle aluminum. If aluminum is worth $1.35 a ton at the recycling center, how much would 15 tons of aluminum be worth? Use estimation to check if your answer is reasonable.

3. The country of Chile in South America produces about 25% (0.25) of the world's copper. How many tons would that be? Use estimation to check if your answer is reasonable.

SKILL TUNE-UP: ADDING AND SUBTRACTING FRACTIONS

In addition to decimals, fractions are also used to represent a part of a whole number, or part of a set.

numerator	$\frac{2}{3}$
denominator	

$\frac{2}{3}$ of the bar above is shaded.

A mixed number is a whole number with a fraction. $1\frac{1}{2}$, $3\frac{4}{7}$

Another way to express a mixed number is with improper fractions. In an improper fraction, the numerator is greater than the denominator.

$$\frac{5}{4}, \frac{14}{5}$$

Equivalent fractions are fractions that stand for the same number. To find equivalent fractions, multiply both the numerator and denominator by the same number.

$$\frac{1}{2} \times \frac{2}{2} = \frac{2}{4} \qquad \frac{1}{2} = \frac{2}{4}$$

$$\frac{1}{3} \times \frac{4}{4} = \frac{4}{12} \qquad \frac{1}{3} = \frac{4}{12}$$

To simplify fractions, divide both the numerator and denominator by their greatest common factor.

$$\frac{3}{9} \div \frac{3}{3} = \frac{1}{3} \qquad \frac{3}{9} = \frac{1}{3}$$

$$\frac{6}{8} \div \frac{2}{2} = \frac{3}{4} \qquad \frac{6}{8} = \frac{3}{4}$$

To add or subtract fractions with different denominators, find the least common multiple of the denominators, then write equivalent fractions for each.

To regroup, use improper fractions.

$$\begin{array}{rcl} 2\frac{1}{4} & = & 1\frac{5}{4} \\ - \ 1\frac{1}{2} & = & 1\frac{2}{4} \\ \hline & & \frac{3}{4} \end{array}$$

$$\begin{array}{rcl} 3\frac{5}{6} & = & 3\frac{15}{18} \\ - \ 1\frac{4}{9} & = & 1\frac{8}{18} \\ \hline & & 4\frac{23}{18} = 5\frac{5}{18} \end{array}$$

A Choose the correct answer.

1. Which of the following fractions is NOT equivalent to $\frac{1}{2}$?

   (A) $\frac{6}{12}$ (B) $\frac{2}{4}$ (C) $\frac{4}{10}$ (D) $\frac{3}{6}$

To compare fractions, find the least common multiple (LCM) of their denominators and rewrite them using the same denominator. Choice C is the only fraction that is not equivalent to $\frac{1}{2}$. Now do the next problem by yourself.

2. $\frac{12}{18}$ is equivalent to—

   (A) $\frac{2}{3}$ (B) $\frac{3}{4}$ (C) $\frac{6}{10}$ (D) $\frac{1}{2}$

3. $\frac{2}{7} + \frac{3}{7} =$

   (A) $\frac{5}{14}$ (B) $\frac{5}{7}$ (C) $\frac{1}{7}$ (D) $\frac{7}{5}$

4. $5\frac{1}{5} - 4\frac{3}{5} =$

   (A) $\frac{1}{5}$ (B) $\frac{4}{5}$ (C) $\frac{3}{5}$ (D) $\frac{2}{5}$

B Solve. Show your work on another piece of paper.

1. Field Day is on May 24. By May 23, Bonnie said that $\frac{9}{12}$ of the class had returned permission slips. When her teacher Ms. Green said she wanted Bonnie to reduce this fraction to simplest form, Bonnie told her it was $\frac{2}{3}$. Was Bonnie correct? Explain your answer.

2. Janice saw an advertisement in the newspaper for a sale on televisions. Electronics City offers a $\frac{1}{3}$ discount off the full price. Television Delight takes $\frac{1}{4}$ off full price. If the full price of the televisions is equal, to which store should Janice go to save the most?

3. A recipe for bread calls for $1\frac{3}{8}$ cups of flour. If Cliff uses $\frac{7}{8}$ of a cup of flour, how much flour is left to add?

STRATEGY: ACT IT OUT

Acting it out is a method that can make adding and subtracting fractions easier. You can use anything from paper clips to classmates to help you act out a story problem.

José had 12 baseball cards. He gave $\frac{1}{4}$ of them to his friend David. How many did he have left?

How could 12 classmates act this out?

$$\frac{12}{12} - \frac{3}{12} = \frac{9}{12}$$

12 cards - 3 cards = 9 cards

Zoe was cutting a pie for her family. She cut $\frac{1}{3}$ of the pie for her parents. She also cut $\frac{1}{2}$ the pie for her brother and sisters. How much of the pie was left for her?

Here's how you could divide a circle to act this out.

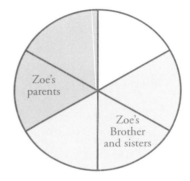

Sam and Jesse were making sandwiches. Sam used $\frac{2}{4}$ of a loaf of bread. Jesse used $\frac{3}{4}$ of his. How much more bread did Jesse use?

Here's how you could use fraction bars to act this out.

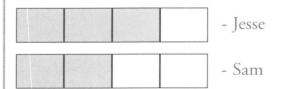

Jesse used $\frac{1}{4}$ more.

A   Choose the best way to use the strategy "act it out."

1. Jerry had taken 6 books to the used book sale. If there were a total of 30 books for sale, find the fraction that Jerry's books represent?

   Ⓐ Draw 30 books and cross out 6 to find there would be 24 books that were not Jerry's.

   Ⓑ Draw 30 books and circle groups of 6 to find his books represent $\frac{1}{5}$ of the total.

   Ⓒ Draw 30 books and add six more books to find his books add to the total.

   Ⓓ Draw 30 books and circle groups of 5 to find his books represent $\frac{1}{5}$ of the total.

   Since there are 30 books, first draw 30 books. Circle groups of 6 books to find there are 5 equal sized groups. Choice B is correct.

B   Solve. Write one or two sentences to describe how to act out each problem.

1. Gloria had 15 gumdrops. She ate $\frac{1}{3}$ of them. How many did she have left?

2. Lonnie brought in 24 cupcakes to share on his birthday. He gave $\frac{1}{6}$ of the cupcakes to teachers and $\frac{5}{6}$ to his classmates. How many cupcakes did his classmates get?

3. On May 20, 1927, Charles Lindbergh flew from New York to Paris in a little over 33 hours. He became the first person to fly nonstop across the Atlantic Ocean. His plane, the Spirit of St. Louis, was tiny by today's standards. It measured about 28 feet in length. If a supersonic jet can fly from New York to Paris in 3 hours, what fraction of Lindbergh's time is this?

L E S S O N 3

STRATEGY: MAKE A TABLE

Organizing information in a table makes it simpler to keep track of numbers, solve problems, and draw conclusions.

You can use a table to list greatest common factors, least common multiples, or equivalent fractions.

Marissa baked her own bread. The recipe called for $\frac{1}{2}$ tablespoon of yeast, but the cookbook also said to decrease it by $\frac{1}{5}$ tablespoon if she was using cake flour. How much yeast should Marissa use along with her cake flour?

$\frac{1}{2} - \frac{1}{5} = ?$

Find the LCM to convert the fractions.

	$\times \frac{1}{1}$	$\times \frac{2}{2}$	$\times \frac{3}{3}$	$\times \frac{4}{4}$	$\times \frac{5}{5}$
$\frac{1}{2}$	$\frac{1}{2}$	$\frac{2}{4}$	$\frac{3}{6}$	$\frac{4}{8}$	$\frac{5}{10}$
$\frac{1}{5}$	$\frac{1}{5}$	$\frac{2}{10}$	$\frac{3}{15}$	$\frac{4}{20}$	$\frac{5}{25}$

$\frac{5}{10} - \frac{2}{10} = \frac{3}{10}$

$\frac{3}{10}$ tablespoon of yeast

A   Choose the letter of the best answer.

1. Which table can help find the simplest form of $\frac{10}{25}$?

Ⓐ
10	20	30	40	50
25	50	75	100	125

Ⓒ
1	5	10	15
25	50	75	100

Ⓑ
1	2	5	10
1	5	25	

Ⓓ
2	4	6	8	10
5	10	15	20	25

To find the simplest form, you need to find the greatest common factor of 10 and 25. Choices A, C, and D do not show this. Choice B is the correct answer. Now try the next one.

2. Which equation can be solved using the following table?

$\frac{7}{8}$	$\frac{7}{8}$	$\frac{14}{16}$	$\frac{21}{24}$	$\frac{28}{32}$
$\frac{5}{6}$	$\frac{5}{6}$	$\frac{10}{12}$	$\frac{15}{18}$	$\frac{20}{24}$

Ⓐ $\frac{7}{8} + \frac{5}{6} = ?$

Ⓑ $\frac{7}{8} \times \frac{5}{6} = ?$

Ⓒ $\frac{5}{6} \div \frac{7}{8} = ?$

Ⓓ $\frac{7}{8} \div \frac{5}{6} = ?$

B   **Solve. Construct a table to help solve the problem, if necessary.**

1. Believe it or not, the continents are drifting apart! They move as much as 1 inch per year. If South America moved $\frac{1}{4}$ inch last year, Europe moved $\frac{3}{8}$ inch and Africa moved $\frac{7}{16}$ inch, construct a table to help find which continent moved the greatest distance.

2. Diane planted a vegetable garden last spring. She planted 24 carrot seeds. By the end of the summer, 18 of the seeds had grown into full-sized carrots. The rest of the seeds did not grow. Express, in simplest form, the fraction of carrot seeds that grew into full-size plants.

3. In some states, many people earn their living in manufacturing. About one-fifth of all working adults have jobs in manufacturing in Massachusetts. Manufacturing employs one-fourth of the working population in Kentucky. Three-tenths of employees in Michigan work in manufacturing. Which state has the largest fraction of its population working in manufacturing?

4. Isabella ran 1 mile Monday, then ran $\frac{1}{4}$ mile more each day through Friday. Shannon ran 2 miles Monday, but only increased her mileage by $\frac{1}{8}$ mile each day. Who ran the most total miles and by how much more?

L E S S O N 4

STRATEGY: LOOK FOR A PATTERN

Finding a pattern can help solve a problem.

Tara increased her babysitting hours every week. Tara started by babysitting for $\frac{3}{4}$ hour the first week. She worked $1\frac{1}{2}$ hours the second, $2\frac{1}{4}$ the third, and 3 the fourth. If the pattern continues, how long will Tara be babysitting after the eighth week?

1	2	3	4	5	6	7	8
$\frac{3}{4}$	$1\frac{1}{2}$	$2\frac{1}{4}$	3	$3\frac{3}{4}$	$4\frac{1}{2}$	$5\frac{1}{4}$	6

The pattern shows that Tara increased her hours by $\frac{3}{4}$ hour each week. She will be babysitting 6 hours after 8 weeks.

Look again at the pattern of Tara's babysitting hours. Notice that not all the fractions have the same denominator because they have been simplified.

When you are trying to determine a pattern, you may find it easiest to compare numbers using equivalent fractions and improper fractions.

Using equivalent fractions, the pattern of Tara's babysitting hours is:

$$\frac{3}{4}, \frac{9}{4}, \frac{12}{4}, \frac{15}{4}, \frac{18}{4}, \frac{21}{4}, \frac{24}{4}$$

 A Choose the letter of the next numbers in the pattern.

1. $\frac{1}{8}, \frac{3}{8}, \frac{2}{8}, \frac{4}{8}, \frac{3}{8}, \ldots\ldots$

   Ⓐ $\frac{5}{8}, \frac{4}{8}, \frac{6}{8}, \frac{5}{8}$ Ⓑ $\frac{5}{8}, \frac{6}{8}, \frac{5}{8}, \frac{7}{8}$ Ⓒ $\frac{5}{8}, \frac{4}{8}, \frac{5}{8}, \frac{7}{8}$ Ⓓ $\frac{2}{8}, \frac{5}{8}, \frac{6}{8}, \frac{5}{8}$

   The pattern is $+\frac{2}{8}, -\frac{1}{8}$. Choices B, C, and D are incorrect. Choice is A is correct. Now find the pattern in the next problem.

2. Ravi kept track of how far a baby snail could go. After the first five minutes, the snail had gone $\frac{1}{12}$ inch. After another five minutes, it had gone $\frac{1}{4}$ inch altogether. The next time Ravi checked, the snail was at the $\frac{5}{12}$-inch mark. Then, the snail moved to the $\frac{7}{12}$-inch mark. At this rate, predict how far the snail will be the next three times Ravi checks its progress.

   Ⓐ $\frac{8}{12}, \frac{3}{4}, \frac{11}{12}$ Ⓑ $\frac{3}{4}, \frac{11}{12}, \frac{13}{12}$ Ⓒ $\frac{7}{12}, \frac{2}{3}, \frac{11}{12}$ Ⓓ $\frac{3}{4}, \frac{5}{6}, \frac{11}{12}$

3. Kelsey sold tickets for a school play. She sold $\frac{1}{2}$ of all her tickets on the first day. The next day she sold $\frac{1}{4}$ of her tickets. On the third day, she only sold $\frac{1}{8}$ of her total tickets. How many will she sell on the next three days?

Ⓐ $\frac{1}{2}$, $\frac{1}{4}$, $\frac{1}{8}$ Ⓑ $\frac{1}{8}$, $\frac{1}{4}$, $\frac{1}{2}$ Ⓒ $\frac{1}{16}$, $\frac{1}{32}$, $\frac{1}{64}$ Ⓓ $\frac{1}{12}$, $\frac{1}{16}$, $\frac{1}{20}$

B   Write a sentence or two to explain your answer.

You know the ocean is salty, but do you realize how salty? If we could remove all the salt from the world's oceans, it would cover the entire earth in a layer 500 feet deep! Imagine filling a container that is 1-foot wide, 1-foot long, and 1-foot high (this is a cubic foot) with seawater. If you leave the container out in the sun so the water evaporates, you would be left with about $2\frac{1}{5}$ pounds of salt.

1. Two containers would leave you with $4\frac{2}{5}$ pounds of salt, and 3 containers would give you $6\frac{3}{5}$ pounds of salt. Describe the pattern you could use if you wanted to find how much salt you would get from 4 containers.

2. If you evaporate salt from a cubic foot of fresh water from a lake or river, you would only be left with $\frac{1}{6}$ of an ounce of salt. Two cubic feet would leave you with $\frac{1}{3}$ of an ounce of salt, and four cubic feet would produce $\frac{2}{3}$ of an ounce of salt. Hunter said that 8 cubic feet of fresh water would leave you with $1\frac{3}{4}$ ounces of salt? Is he correct?

Do you like red or green ketchup? Maybe you'd prefer purple? Many companies have found that kids love different colored food. Blue applesauce, pink margarine, and blue French fries are just some of the products being offered.

3. Some experts are concerned that the popularity of the colored foods is part of a larger problem. Each year, children are eating more and more junk food. Right now, it is estimated that $\frac{1}{5}$ of all American children need to lose weight. Health officials would like to decrease this number through education. They would be happy with the results shown in the table below. If the pattern continues, what fraction of the children will be overweight in the 5th and 6th years?

Year	1	2	3	4
Fraction Overweight	$\frac{1}{5}$	$\frac{9}{50}$	$\frac{4}{25}$	$\frac{7}{50}$

CHOOSE A STRATEGY

Review the strategies you have learned so far.

Act It Out—Acting out a problem can make it easier. Use anything from paper clips to classmates to help solve a story problem.

Make a Table—Organizing information in a table makes it simpler to keep track of information, to solve problems, and to draw conclusions.

Candice and Jada were painting a fence. After the first hour, Candice had done $1\frac{1}{2}$ ft and Jada had completed $1\frac{1}{3}$ ft. If each girl continued at that rate, how many feet would they have completed altogether after 4 hours?

Hour	Candice	Jada
1	$1\frac{1}{2}$ ft	$1\frac{1}{3}$ ft
2	3 ft	$2\frac{2}{3}$ ft
3	$4\frac{1}{2}$ ft	4 ft
4	6 ft	$5\frac{1}{3}$ ft

6 ft $+ 5\frac{1}{3}$ ft $= 11\frac{1}{3}$ ft altogether.

Look for a Pattern—Finding a pattern can help solve a problem.

Garrett biked 11 miles before resting. Then, he biked 10 miles more before resting again, then 8 miles before another rest. The pattern is to subtract 1 mile more than was subtracted before. If his fifth stop is the end of the trail, how far did Garrett travel?

The pattern would be 11, 10, 8, 5, 1

Add the miles together to find that Garrett biked 35 miles altogether.

A   Choose the letter of the correct answer.

1. Paul and Neve studied hard for their math test. On Tuesday, Paul studied $1\frac{1}{4}$ hours and Neve studied $1\frac{1}{2}$ hours. On Wednesday, Neve studied for $1\frac{1}{3}$ hours and Paul studied for $1\frac{1}{2}$ hours. To find who studied the most on Tuesday and Wednesday, which strategy would you choose?

   Ⓐ solve a simpler problem
   Ⓑ make a table
   Ⓒ look for a pattern
   Ⓓ act it out

B Write a sentence or two to explain your answer to the question.

1. Many children grow rapidly in their teen years. Last year, Esteban grew $\frac{1}{4}$ inch in January, $\frac{1}{2}$ inch in March, and $\frac{3}{4}$ inch in May. If this trend continues, how would you find how many inches Esteban will grow by September?

2. The Tigers soccer team plays a 14-game schedule. They lost their first 4 games and won half of their remaining games. What strategy would help to find their final won-lost record?

C Solve. Show your work on another sheet of paper.

1. Until 10 years ago, charcoal grills were far more popular than gas grills. Some supergrills cost over $1,000. Most dealers sell these along with lower cost models. In one dealer's showroom there are 12 grills. $\frac{1}{6}$ of the grills are gas supergrills, $\frac{1}{3}$ are lower price charcoal, $\frac{1}{4}$ are lower price gas, and 3 are electric supergrills. List the grills being displayed from least to greatest number.

2. People are eating less red meat. If $\frac{3}{4}$ of all grills were used to grill red meat in 1990, $\frac{7}{12}$ grilled red meat in 1995, and $\frac{5}{12}$ grilled red meat in 2000, what fraction of grills will be used for red meat in 2005 if this trend continues?

3. Usually, Mrs. Moore is in charge of cooking dinner. In the summertime, however, it is a different story—Mr. Moore often takes over on the grill. During June, Mr. Moore grills chicken $\frac{1}{3}$ of the time, hamburgers $\frac{1}{6}$, fish or shrimp $\frac{2}{15}$, and steaks $\frac{1}{30}$ of the days. The rest of the time, Mrs. Moore cooks. How many days is Mrs. Moore in charge of dinner? (Hint: June has 30 days)

4. Raul's electric grill has space for 20 drumsticks or hot dogs. Raul cooked 4 shish kabobs, which took up $1\frac{1}{2}$ spaces apiece. He used half of the spaces left for drumsticks. How many hot dogs could Raul cook?

L E S S O N   6

MULTI-STEP STORY PROBLEMS

 A Circle the letter of the correct answer.

1. Sondra ran $\frac{1}{2}$ mile on Friday, $\frac{1}{4}$ mile on Saturday, and $\frac{3}{8}$ of a mile on Sunday. Karla ran $\frac{1}{5}$ mile on Friday, $\frac{1}{4}$ mile on Saturday, and $\frac{3}{10}$ mile on Sunday. Which table can be used to find who ran a greater total distance?

Ⓐ

Name	Friday	Saturday	Sunday
Sondra	$\frac{1}{2}$	$\frac{1}{4}$	$\frac{3}{10}$
Karla	$\frac{1}{5}$	$\frac{1}{4}$	$\frac{3}{8}$

Ⓒ

Name	Friday	Saturday	Sunday
Sondra	$\frac{1}{2}$	$\frac{1}{4}$	$\frac{3}{8}$
Karla	$\frac{1}{5}$	$\frac{1}{4}$	$\frac{3}{10}$

Ⓑ

Name	Friday	Saturday	Sunday
Sondra	$\frac{1}{4}$	$\frac{1}{4}$	$\frac{3}{8}$
Karla		$\frac{1}{4}$	$\frac{3}{10}$

Ⓓ

Name	Friday	Saturday	Sunday
Sondra	$\frac{1}{2}$	$\frac{1}{5}$	$\frac{3}{8}$
Karla	$\frac{1}{5}$	$\frac{1}{4}$	$\frac{3}{10}$

2. In the first year, $\frac{1}{8}$ of all adults voted in an election. After a registration drive the second year, $\frac{1}{4}$ voted. The following year, $\frac{1}{2}$ of all adults voted. If $\frac{7}{8}$ of adults voted the next year, what is the pattern?

Ⓐ $+\frac{1}{8}, +\frac{3}{8}, +\frac{2}{8}$

Ⓒ $+\frac{3}{8}, +\frac{2}{8}, +\frac{1}{8}$

Ⓑ $+\frac{1}{8}, +\frac{2}{8}, +\frac{4}{8}$

Ⓓ $+\frac{1}{8}, +\frac{2}{8}, +\frac{3}{8}$

B Write a sentence or two to explain your answer to the question.

1. Many types of rock are used in building construction. Sandstone, limestone, granite, and marble are dug from quarries and cut into smaller pieces before being shipped. Of 30 pieces of building material, $\frac{1}{5}$ are granite, $\frac{1}{6}$ marble, $\frac{1}{3}$ sandstone, and the remainder is limestone. How could you find the number of pieces of limestone?

2. Bricks are made mostly from clay. The clay is ground into powder and then mixed with water and other materials. It is then placed in molds and dried under high heat. At the factory, $\frac{1}{16}$ of the bricks are dried in the first day. The second day, $\frac{3}{32}$ of the bricks are dried and the third day, $\frac{1}{8}$ of the bricks are dried. How could you find the fraction of bricks that will be dried the fourth day?

C   Solve the problem. Show your work on another sheet of paper.

1. Hardware screws come in many lengths. Their length ranges from $\frac{1}{4}$ inch long to over 5 inches. It is important to use the proper size screw for the job you are doing. To fasten a half-inch piece of wood, Bruce chose a $\frac{1}{4}$ -inch screw. To fasten a one-inch piece of the same material, he chose a half-inch screw. A $\frac{3}{4}$ -inch screw was used to fasten a 1 $\frac{1}{2}$ -inch piece of wood. What length screw will Bruce use to fasten a 2-inch piece of wood?

2. To do a job correctly, it is important to cut lumber to exact lengths. Alex will be using pieces of wood 2 $\frac{1}{8}$ feet, 3 $\frac{1}{2}$ feet, and 4 $\frac{1}{6}$ feet in length. Cathy is using pieces of wood 2 $\frac{1}{4}$ feet, 3 $\frac{3}{4}$ feet, and 4 $\frac{1}{4}$ feet long. Who will use more lumber?

3. When T.J. moved away, his friend Jonah gave him a small tree. Jonah told T.J that they would have a contest to see which tree would grow the fastest. In one week, T.J.'s tree usually grew $\frac{1}{2}$ inch. Jonah's tree usually grew $\frac{3}{4}$ inch. But when either would have a rainy week, the tree would grow $\frac{1}{2}$ inch more than usual. In 6 weeks, T.J. had 4 rainy weeks, while Jonah had one. Whose tree grew the tallest?

4. Leilani bought a CD carrier so that she could take 36 CDs with her wherever she went. Leilani put hip-hop CDs in $\frac{1}{3}$ of the slots. Then, she put movie soundtracks in $\frac{1}{6}$ of the remaining slots. Next, she put country CDs in $\frac{1}{2}$ half of the remaining slots. How many slots does she have for her rock CDs?

SKILL TUNE-UP: MULTIPLYING AND DIVIDING FRACTIONS

To multiply fractions, multiply numerator by numerator and denominator by denominator.

$$\frac{2}{3} \times \frac{2}{5} = \frac{2 \times 2}{3 \times 5} = \frac{4}{15}$$

To divide fractions, multiply the dividend by the reciprocal of the divisor.

$$\frac{4}{5} \div \frac{1}{2} \rightarrow \frac{4}{5} \times \frac{2}{1}$$

To multiply or divide whole or mixed numbers by a fraction, first change them into improper fractions.

$$6 \times \frac{3}{5} \rightarrow \frac{6}{1} \times \frac{3}{5}$$

$$2\frac{1}{2} \div \frac{1}{3} \rightarrow \frac{5}{2} \div \frac{1}{3} \rightarrow \frac{5}{2} \times \frac{3}{1}$$

To make a problem easier, look for common factors between numerators and denominators. Divide each by a common factor, then use the answer in their place when multiplying or dividing.

$$\frac{4}{9} \times \frac{6}{8} \qquad \frac{4}{9} \times \frac{\overset{3}{6}}{\underset{4}{8}} \qquad \frac{4}{9} \times \frac{\overset{1}{3}}{\underset{1}{4}}$$

Remember to simplify your answer and change improper fractions back into mixed numbers.

$$\frac{1}{9} \times \frac{3}{1} = \frac{3}{9} = \frac{1}{3}$$

 A Choose the correct answer.

1. $\frac{1}{7} \times \frac{2}{3} =$

   Ⓐ $\frac{3}{14}$ Ⓑ $\frac{21}{2}$ Ⓒ $\frac{14}{3}$ Ⓓ $\frac{2}{21}$

Choice D is the correct answer. Now try division.

2. $\frac{1}{7} \div \frac{2}{3} =$

   Ⓐ $\frac{3}{14}$ Ⓑ $\frac{21}{2}$ Ⓒ $\frac{14}{3}$ Ⓓ $\frac{2}{21}$

3. $\frac{3}{4} \times \frac{1}{2} =$

   Ⓐ $\frac{3}{8}$ 　　　　Ⓑ $1\frac{1}{2}$ 　　　　Ⓒ $\frac{2}{3}$ 　　　　Ⓓ $1\frac{1}{4}$

4. $2\frac{1}{3} \times 3\frac{1}{2} =$

   Ⓐ $\frac{2}{3}$ 　　　　Ⓑ $8\frac{1}{2}$ 　　　　Ⓒ $8\frac{1}{6}$ 　　　　Ⓓ $\frac{3}{4}$

 Write an equation or a sentence to solve the problem.

1. Dana walked $\frac{1}{2}$ mile to and from soccer practice 4 days each week. Dana said she walked $1\frac{1}{2}$ miles each week. Is Dana correct? Explain your answer.

2. The number of horseshoe crabs on the East Coast is decreasing. About $11\frac{1}{2}$ million horseshoe crabs were counted during a survey taken between Cape May, New Jersey and Ocean City, Maryland. This may sound like a lot, but it is far less than past estimates. The actual number may be as little as $\frac{2}{3}$ this count. Write the equation to show that horseshoe crabs may number as low as $7\frac{2}{3}$ million.

3. The section surveyed for the horseshoe crab census is only about 1/40 of the approximately 2,000-mile East Coast. About how long is the section that was surveyed?

4. Arnie baked a cake for his mother's birthday party. He served $\frac{3}{4}$ of the cake. Each party guest ate $\frac{1}{8}$ of the cake that was served. How many people were at the party?

5. Jeremy decided to study a total of $2\frac{1}{2}$ hours for his Math final exam. On Monday, he studied for half the total time he set aside for studying. How long has Jeremy studied so far?

STRATEGY: WRITE AN EQUATION

Writing an equation can help you find the important information and determine the operations to use.

Tawana is using $\frac{5}{8}$ cup of sugar for her cookie recipe. If a recipe for brownies calls for $\frac{1}{3}$ cup more sugar than is needed for cookies, how much sugar is needed for the brownie recipe?

What we know Cookies need $\frac{5}{8}$ cup of sugar. Brownies need that amount and $\frac{1}{3}$ more.

Our plan Write an equation to help solve the problem.

Calculate

$\frac{5}{8} + (\frac{5}{8} \times \frac{1}{3}) = s$

$\frac{5}{8} + (\frac{5}{24}) = s$

$\frac{15}{24} + \frac{5}{24} = s$

$\frac{20}{24} = s \quad = \frac{5}{6}$ cup of sugar

Remember: whatever you do to one side of the equation, you must do to the other side as well.

Tawana only added $\frac{3}{8}$ cup of sugar to her cookie dough. How much more does she need to add to get the correct amount of $\frac{5}{8}$ cup sugar?

Calculate

$\frac{3}{8} + m = \frac{5}{8}$

$(\frac{3}{8} - \frac{3}{8}) + m = \frac{5}{8} - \frac{3}{8}$

$m = \frac{2}{8} \qquad m = \frac{1}{4}$

Tawana needs to add $\frac{1}{4}$ cup more sugar.

A Choose the letter of the correct answer.

1. Rachel and Zach painted their rooms. Rachel spent $\frac{1}{3}$ hour longer painting her room than Zach. Zach spent $3\frac{1}{2}$ hours painting his room. Which equation shows how long Rachel spent painting her room?

   Ⓐ $n + 3\frac{1}{2} = \frac{1}{3}$ 　　Ⓑ $3\frac{1}{2} - \frac{1}{3} = n$ 　　Ⓒ $n + \frac{1}{3} = 3\frac{1}{2}$ 　　Ⓓ $n = 3\frac{1}{2} + \frac{1}{3}$

   The solution to choice A is a negative number. In B and C, each equation would show that Rachel worked $\frac{1}{3}$ less— not more. Choice D is the correct equation. Find the next equation on your own.

2. Kevin bought a package of paper with 100 sheets to use for writing reports. He plans to use $\frac{3}{4}$ of the sheets and save the rest for scrap paper. If he uses $\frac{1}{3}$ of the sheets saved for reports in the first week, which equation can be used to find the number of sheets of paper Kevin used?

Ⓐ $(100 \times \frac{1}{3}) + \frac{3}{4} = n$

Ⓒ $(\frac{3}{4} \times \frac{1}{3}) + 100 = n$

Ⓑ $(100 \times \frac{3}{4}) \times \frac{1}{3} = n$

Ⓓ $(\frac{1}{3} \times \frac{3}{4}) \div n = 100$

B Solve. Show your work on another sheet of paper.

A popular pet is a small lizard called a Bearded Dragon. A flap of skin under the jaw turns black when they get excited. They never get too large (about 2 feet long) and they are easy to handle. Beardies, as they are known, will eat plants and insects. I hope you have the chance to care for one of these fascinating animals!

1. It is important to maintain a certain range of temperatures for Beardies. The minimum temperature should be around 80 degrees. The maximum temperature these lizards can tolerate is $1\frac{1}{6}$ times the minimum. Write an equation to find the highest temperature they could be exposed to.

2. Bearded lizards need food dusted with calcium and vitamin D3 twice a week. Kay put 4 pounds of calcium tablets into one-quarter pound containers so she can keep them fresh. She also put 2 pounds of vitamin D3 capsules into one-quarter pound containers. How many containers did Kay need?

Francis Scott Key was so inspired by the flag flying over Fort McHenry during a fierce battle in 1813 that he wrote the USA national anthem, "The Star Spangled Banner." Sadly, the flag he saw is in trouble. Bullets and souvenir seekers have left holes in it. Experts are now at work fixing the flag at the National Museum of American History in Washington, D.C.

3. The flag is about 34 feet long. This may seem impressive, but it is only $\frac{4}{5}$ the original length. Using an equation, what was the length of the flag in 1813?

4. Experts work with tiny surgical tools to repair the flag. The work is so difficult, that for every 45 minutes ($\frac{3}{4}$ hour) they work, the conservators must rest 15 minutes ($\frac{1}{4}$ hour). If Mike worked 6 hours, write two equations to find how much time he spent working and how much time he spent on breaks.

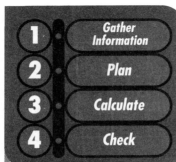

L E S S O N 3

STRATEGY: DRAW A PICTURE

A picture can sometimes help you solve a problem.

After dinner, Lalo had $\frac{3}{4}$ of a pizza left. If he wanted to share the rest with 5 friends, how much of the entire pizza would each person eat?

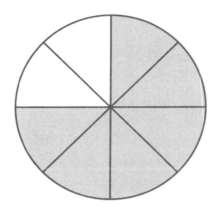

Lalo and his friends will each eat $\frac{1}{8}$ of the pizza.

 Choose the best answer.

1. Valerie is reading a book for her history report that has 10 chapters. So far, Valerie has read $\frac{8}{10}$ of the book. She can read 2 chapters a day. Which picture can help find how many days Valerie has been reading her book?

Ⓐ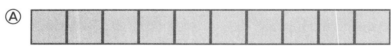

Ⓑ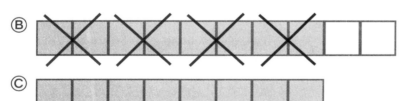

Ⓒ

Ⓓ

Choice B is the best answer. It takes Valerie 4 days. Find the next correct diagram by yourself.

2. Jeffrey wants to save $15 from his allowance for a CD. So far, he has saved one-fifth that amount. Which picture can be used to find how much Jeffrey has saved so far?

Ⓐ $ $ $ $ $
   $ $ $ $ $
   $ $ $ $ $

Ⓒ $ $ $ $
   $ $ $ $
   $ $ $ $
   $ $ $ $

Ⓑ $ $ $ $ $
   $ $ $ $ $
   $ $ $ $ $
   $ $ $ $ $
   $ $ $ $ $

Ⓓ $ $ $
   $ $ $
   $ $ $
   $ $ $
   $ $ $

B **Write a sentence or draw a picture to answer the question.**

1. Tara had Math, Social Studies, Science, and Language Arts homework, each taking the same amount of time. After half an hour, she had completed half her Science assignment. Explain why the diagram below can be used to find what fraction of her total homework assignment Tara has completed.

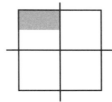

2. Omar built some bookshelves for his room using a 10-foot piece of lumber. Omar cut $2\frac{1}{2}$ -foot pieces for each shelf. Draw a picture to help find how many bookshelves Omar can make from the single piece of lumber.

   In the 1930s, a fur-bearing rodent from South America, the nutria, was introduced into the swamps of Louisiana. At the time, the nutrias' fur could be sold for large sums of money. During the 1980s, many people decided they did not want to wear fur.

3. As a result of the decrease in people wearing fur, the number of trappers was also reduced. At the peak in demand, there were 12,000 trappers. Currently, there are about $\frac{1}{12}$ this number. How many trappers are there today?

CHOOSE A STRATEGY

Review the strategies you have learned so far.

Write an Equation—Writing a number sentence can help you decide which numbers and operations to use.

Grandin knew that he needed $\frac{5}{8}$ yard of plywood to make a birdhouse. His mother asked him to make 3 birdhouses. How much plywood does he need?

$\frac{5}{8} \times 3 = w$ $\frac{5}{8} \times \frac{3}{1} = w$

$\frac{15}{8} = w$ $1\frac{7}{8} = w$

Grandin needs $1\frac{7}{8}$ yards of plywood.

Draw a Picture—Drawing a picture to represent a problem may help solve it.

Grandin found $\frac{3}{4}$ of a can of paint in the garage. His mother told him that she needed $\frac{1}{2}$ of the paint left for her own projects. How much of the original can of paint could Grandin use for birdhouses?

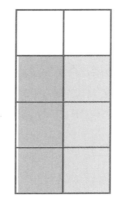

Grandin could use $\frac{3}{8}$ of the paint.

A Choose the correct answer.

1. Of 400 people at the softball game, $\frac{1}{4}$ of them bought a hot dog and soda. Which equation could be used to find how many people did not buy a hot dog and soda?

   Ⓐ $400 - (400 \times \frac{1}{4}) = n$ Ⓒ $400 + (400 \times \frac{1}{4}) = n$

   Ⓑ $400 - (400 \div \frac{1}{4}) = n$ Ⓓ $400 + (400 \div \frac{1}{4}) = n$

B Write an equation or a sentence to solve the problem. Show your work.

Many people love to plant rhododendrons. They come in all sizes from dwarf varieties to giants. They come in a wide range of colors—from white to red to yellow. They are a beautiful addition to any garden.

1. Some rhododendrons can be quite expensive. One type, Mary Bell, has beautiful peach-colored flowers. It sells for $20.00. Another variety, Vulcan, has bright red flowers. It costs $1\frac{3}{4}$ as much as Mary Bell. Write the equation that can be used to find the cost of the Vulcan rhododendron.

2. The Tapestry rhododendron grows to a height of 4 feet. This may seem high, but it is only $\frac{2}{3}$ the height of the Taurus variety. Bob used the following equation: $4 \times \frac{2}{3} = h$ to find the height of the Taurus rhododendron. Did Bob use the correct equation? Explain your answer.

If you love to hike, rollerblade, or bike ride, then New York's Hudson Valley is the place for you. Lengths range from the 1.3-mile Jones Point Greenway Trail to the 124-mile Hudson Valley Greenway Bike Route. There is a trail or path to suit everyone.

3. Christopher can bike up to 10 miles per hour. His friend Alice has trained more and is faster. Christopher's speed is only $\frac{3}{4}$ the speed of Alice. How fast can Alice bike?

4. The Wilbur Boulevard Trailway is paved and runs for $1\frac{1}{5}$ miles. If Paige has completed $\frac{1}{4}$ of the trail, how far has she walked?

MULTI-STEP STORY PROBLEMS

A Choose the correct answer.

1. Ben needs to tie packages with ribbon. He has 3 yards of ribbon and each package needs $\frac{1}{3}$ yard of ribbon. Which equation could be used to find out if he has enough for 10 packages?

　Ⓐ $10 \times 3 = ?$　　　　　　　　　　Ⓒ $3 \div \frac{1}{3} = ?$

　Ⓑ $10 \div 3 = ?$　　　　　　　　　　Ⓓ $3 \times \frac{1}{3} = ?$

2. Karen has 15 birthday cards. She will send $\frac{1}{3}$ of the cards to her aunts and $\frac{1}{5}$ to her cousins. Which equation can be used to find how many birthday cards she will not send?

　Ⓐ $15 + (15 \times \frac{1}{3}) - (15 \times \frac{1}{5}) = n$　　　　Ⓒ $15 - (15 \times \frac{1}{5}) - (15 \times \frac{1}{5}) = n$

　Ⓑ $15 - (15 \div \frac{1}{3}) - (15 \times \frac{1}{5}) = n$　　　　Ⓓ $15 - (15 \times \frac{1}{3}) - (15 \times \frac{1}{5}) = n$

3. One-fourth of Ms. Murthi's class is in the talent show. Of the students in the talent show, half will sing. If there are 32 students in Ms. Murthi's class, how many will sing?

　Ⓐ 4　　　　　　　　　　　　　　Ⓒ 16

　Ⓑ 8　　　　　　　　　　　　　　Ⓓ 32

B Draw a picture or write an equation to solve the problem. Show your work.

When most people think of desert wildlife, they think of snakes, lizards, and scorpions. But the desert bee is also an important part of the desert ecosystem. This bee pollinates the flowers of many desert plants and trees.

1. Compared to honeybees, the desert bee is tiny. Most desert bees average just one-third inch in length. This is one-half the length of honeybees. Write an equation that would help find the combined length of 5 honeybees.

2. Desert bees gather in large colonies. They may do this to protect themselves from their enemy, the assassin bug. This predator kills by stabbing its prey with its beak and injecting a poison. In one colony, $\frac{2}{3}$ of the bees may be protected. Perhaps $\frac{1}{6}$ of the unprotected bees will be killed by assassin bugs. What fraction of the colony may be destroyed?

3. During a windstorm, the bees take shelter in the hive. A third of the bees rest in the top third of the hive. If the wind blows the top third off, and $\frac{1}{3}$ of the bees in it do not escape, what fraction of the beehive goes with the top?

About 300 sunken ships have been found in the Gulf of Mexico in recent years. Only one-third of these vessels have been identified. Using modern technology, it is hoped the identity of more of the ships will become known.

4. If the fraction of ships identified increases to one-half, how many more ships will be identified compared to what is known today?

5. When examining wrecks 2,000 feet below the surface, a remote control submarine is sent to take pictures. For more shallow wrecks, a 145-foot submarine with a crew of 5 is used. This submarine is much larger than many of the ships it is exploring. A sixteenth-century ship recently discovered was only about $\frac{3}{10}$ as long as the submarine. What was the length of the wreck?

SKILL TUNE-UP: RATIO, PERCENT, AND PROBABILITY

A ratio compares 2 quantities.

The ratio of the number of heads on a coin to the number of tails is 1:1. The ratio of heads to sides is 1:2.

A percent compares a number to 100.

There is a 50% chance of flipping a head on a coin. In other words, if you flipped the coin 100 times, the head is likely to appear 50 times.

A 100% chance exists for flipping a head OR a tail on a coin.

Probability is the chance that an event will happen.

$$\frac{\text{The desired outcome (heads)}}{\text{Number of possible outcomes (heads or tails)}} = \frac{1}{2}$$

Since probability may be expressed as a percent, fraction, or decimal, the probability of the coin landing on heads may be written as $\frac{1}{2}$, 50%, or 0.50.

Just like fractions, a probability should be simplified to its lowest term.

A Choose the best answer.

1. Which of the following is **not** equivalent to 6:2?

   Ⓐ 3:1   Ⓑ 2:6   Ⓒ $\frac{6}{2}$ Ⓓ 60:20

   Choice A is the same as 6:2 in its simplest form. Choice C is another way to express the same probability. Choice D, when reduced, is the same as 6:2. The correct choice is B, which is **not** equivalent to 6:2. Find the next probability by yourself.

2. Using a spinner with equal sections numbered 1-4, what is the probability the spinner will land on 3?

   Ⓐ 25%   Ⓑ 4%   Ⓒ 40%   Ⓓ 20%

3. Dan reached into a bag with 6 red marbles and 4 blue marbles. What is the probability of Dan picking a blue marble?

   Ⓐ $\frac{2}{3}$ Ⓑ 50%   Ⓒ 0.60   Ⓓ 2:5

4. What is the probability of rolling an even number when rolling a die numbered 1-6?

Ⓐ 1:6   Ⓑ $\frac{1}{3}$ Ⓒ 0.50   Ⓓ 60%

 Write a phrase or sentence to answer the question.

1. Fran is conducting a probability experiment. She will be tossing a coin 100 times. She predicts that the probability it will land on heads is 100%. Why is Fran wrong?

2. There is a special contest at the movie theater on Saturday night. For every 100 customers, 10 will win a free pass to the movies. Jim says there is a 10% chance of winning a free pass. How does Jim calculate this probability?

3. In the 2002 World Cup Soccer Championship, 32 teams from around the world were equally divided into 8 groups lettered A – H. What is the probability the United States team had been placed in group D?

 Solve. Show your work.

1. There are 15 girls in a class of 25 students. Each student's name is written separately on an index card, and the 25 cards are placed in a box. What is the probability of reaching in and picking a card with the name of a boy?

2. For homework, Miss Fryman gave each of her students a blank 8-sectioned spinner. She told them to number each section with a 1 or a 2 so there is a 3:4 probability the spinner will not land on an even number. How can the spinner be numbered? Explain your answer.

STRATEGY: MAKE A TALLY CHART

Making a tally chart can help you solve a probability problem.

Alex was playing a board game with a six-sided cube, numbered from 1 to 6. What is the probability of the number cube landing on an odd number when Alex rolls it?

First, list all the possibilities in columns. Next, perform the experiment, and tally each outcome.

Alex landed on a one, three, or five 10 times out of the 20 times he rolled the cube.

The probability is $\frac{1}{2}$, 0.5, or 50%.

Your experiments may not match exactly your prediction of probability. Do the experiment as many times as you can to get the most realistic probability.

I	2	3	4	5	6
卌	IIII	III	IIII	II	II

A Choose the correct answer.

1. Paul tossed a coin 10 times. It landed on heads 5 times and tails 5 times. Which tally chart shows Paul's results?

Ⓐ
H	T
卌	卌

Ⓒ
H	T
卌	IIII

Ⓑ
H	T
卌 I	IIII

Ⓓ
H	T
卌 卌	

The tally chart must reflect the results of the experiment. Choice B only shows the coin landing on tails 4 times, while C shows only 9 tosses. Choice D shows all 10 coin tosses landing on heads. Choice A correctly shows the result of the experiment. Now, try the next one on your own.

Write a phrase or sentence to answer the question.

1. Chelsey has a bag with 5 green, 3 blue, and 4 red crayons. When she reaches in to pick a crayon, she records the results and puts the crayon back in the bag. On her picks, she selects blue, blue, green, red, and green crayons. Draw a tally chart to show these results.

2. If you toss a coin, what is the probability it will land on heads? What is the chance it will land on tails? Toss a coin 50 times and record your results in a tally chart. How does this compare with your prediction?

 Solve. Show your work.

1. Write the name of each day of the week on its own slip of paper, then turn all the slips upside down and mix them up. What do you think the probability is of picking a weekend day? Now, reach in and pick a slip 14 times, replacing the slip after each selection. Keep a tally chart of the days you pick. Did the results match your expectations? Explain your answer.

2. There is a spinner with 6 sections marked A, B, C, A, D, and E. What is the probability of:

   The spinner landing on A?

   It not landing on C?

   It landing on F?

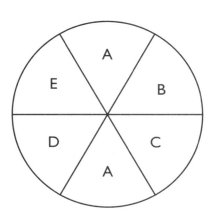

STRATEGY: USE PROPORTIONS

A proportion is a number sentence which uses equivalent ratios.

For one serving of oatmeal, use 4 tablespoons of oats. For 8 servings of oatmeal, use 32 tablespoons of oats.

$$\frac{\text{serving size}}{\text{amount of oats}} = \frac{1}{4} = \frac{8}{32}$$

Use cross multiplication to find a missing number in a proportion.

In Brianna's bowl of fruit, the ratio of cherries to grapes is 6 to 9. If she had 30 cherries, how many grapes would Brianna have?

1) Write each ratio as a fraction. Do not reduce fractions until you have the answer.

$$\frac{6}{9} = \frac{30}{?} \qquad\qquad \frac{6}{9} \diagdown\!\!= \frac{30}{?}$$

2) Cross-multiply to solve the equation.

6 x ? = 9 x 30
6 x ? = 270
? ÷ 6 = 270 ÷ 6   ? = 45 grapes

Use proportions to find percentages. When finding percentages, you can reduce fractions to make problems easier to solve.

Emilio figured out that 25% of the houses on his newspaper route have dogs. If Emilio delivers papers to 32 houses, how many of them have dogs?

$$\frac{25}{100} = \frac{?}{32} \longrightarrow \frac{1}{4} = \frac{?}{32}$$

4 x ? = 1 x 32
4 x ? = 32
? ÷ 4 = 32 ÷ 4
? = 8 houses have dogs

A Choose the missing number.

1. $\dfrac{5}{7} = \dfrac{10}{?}$

   Ⓐ 12　　　　　　　Ⓑ 16　　　　　　　Ⓒ 18　　　　　　　Ⓓ 14

   To find the missing number in a proportion, first cross-multiply 7 x 10 = 70. Divide by 5 to find the missing numerator of 14. Choice D is the correct answer.

2. $40\% = \dfrac{?}{20}$

   Ⓐ 8　　　　　　　Ⓑ 10　　　　　　　Ⓒ 6　　　　　　　Ⓓ 12

B Write a phrase or sentence to answer the question.

1. A map of New York State is drawn using a scale of 1 inch = 20 miles. When Joan measures Long Island on the map, it measures 5 inches in length. She calculated that Long Island is 100 miles long. How do you know Joan is correct?

2. Clark learned that many plants produce a great number of seeds to ensure their survival. Only a small percentage actually sprout. The common dandelion produces hundreds of seeds, and yet only 10% may produce new plants. Clark used the following proportion to predict how many of 200 dandelion seeds may sprout:

   $\dfrac{200}{10} = \dfrac{100}{?}$

   Clark's Science teacher told him his proportion wasn't correct. How can Clark set up the proper proportion to make his prediction?

3. Think of your fingernails as nature's armor. They protect the soft flesh on our fingers and they help us scratch insect bites. Without fingernails we would have a hard time picking up tiny objects. Fingernails grow faster in summer than winter. They grow at a rate of $\dfrac{1}{8}$ inch per month. At this rate, how many inches will fingernails grow in a year?

L E S S O N 4

CHOOSE A STRATEGY

Review the strategies you have learned so far.

Make a Tally Chart— List all the possibilities in a chart, then use the chart to tally an experiment's outcome.

Mauli put 2 pennies and 2 nickels in a bag. When blindly picking 2 coins at the same time, what is the probability of her choosing 2 nickels?

1 penny 1 nickel	2 pennies	2 nickels
I	I	I

The probability of choosing 2 nickels is $\frac{1}{3}$.

Proportions— Use cross multiplication to find a missing number in a proportion.

In 2 packs of candy, 8 pieces are mints. How many packs do you need in order to have 20 mints?

$$\frac{2}{8} = \frac{?}{20}$$

8 x ? = 2 x 20
8 x ? = 40
? = 40 ÷ 8
? = 5 packs of candy

A Choose the letter of the correct answer.

1. A basketball player made **60%** of her free throws. If she shot **80** times, how many went in the basket?

   Ⓐ 60   Ⓑ 58   Ⓒ 48   Ⓓ 38

To find the number of free throws made, set up a proportion: $\frac{60}{100} = \frac{?}{80}$

Cross multiply 60 x 80 = 4,800 and divide 4,800 ÷ 100 = 48. The correct choice is C.

2. **40 pennies were accidentally dropped on the floor. What is the probability a penny lands heads up?**

   Ⓐ 1:2   Ⓑ 2:1   Ⓒ $\frac{1}{40}$ Ⓓ 20 to 1

3. The tally chart below shows the results of a class poll on favorite ice cream flavors. Based on these results, how many people can be expected to name vanilla as their favorite if 48 people are surveyed?

Chocolate	Vanilla
~~IIII~~ III	IIII

Ⓐ 12   Ⓑ 16   Ⓒ 32   Ⓓ 8

B Solve. Show your work on another piece of paper.

1. Al was conducting a study of cars going over a bridge during rush hour. Of the cars he counted, 160 had 2 passengers, 200 had only the driver, and 140 had more than 2 passengers. The next day, 750 cars drove over the bridge. How many had only one driver?

2. In a can of assorted nuts, there are 100 peanuts, 50 pecans, 75 cashews, and 75 macadamia nuts. Janet wanted to find the probability of not picking a peanut if she reached in without looking. Explain how she can find the chance of this happening.

3. Sound travels faster in warm air. It travels at a rate of 331 meters per second in air at 0 degrees Celsius and 343 meters per second at 20 degrees. When air is warmed to 25 degrees, sound travels at a rate of 346 meters per second. Based on this information, complete the chart below. Show your work.

Temperature	0 degrees Celsius	20 degrees Celsius	25 degrees Celsius
Distance	993 meters		
Time		5 seconds	$3\frac{1}{2}$ seconds

4. Everyone wants to be an actor, but it is not an easy way to make a living. For every star, there are many people who have tried without success. At auditions, hundreds of people may try out for a single part. At one audition, 120 people tried out for 12 parts. What is the probability of someone getting a part? What percentage of people were successful?

MULTI-STEP STORY PROBLEMS

A Circle the letter of the correct answer

1. Loki bought 4 muffins for $3. What percent of a dollar does each muffin cost?

   Ⓐ 85%   Ⓑ 45%   Ⓒ 30%   Ⓓ 75%

2. Lauren tried a probability experiment with a six-sided number cube. After 10 tosses, Lauren recorded these results:

1	2	3	4	5	6
II	I	I	I	II	III

   Which of the following statements is true?

   Ⓐ There were more even tosses than odd tosses.

   Ⓑ There were an equal number of odd and even tosses.

   Ⓒ The cube proved there was a $\frac{1}{6}$ chance for any number.

   Ⓓ The next toss of the cube will probably land on 6.

3. Which of the following would have the greatest probability?

   Ⓐ A penny landing on heads or tails.

   Ⓑ A four-sectioned spinner landing on 2.

   Ⓒ A six-sided number cube landing on 3.

   Ⓓ A person having a birthday in July.

4. On his last Math test, Larry got 80% of the questions correct. Before the next test, he studied each night and scored 90%. If each test had 30 questions, how many questions did Larry answer correctly on his second test?

   Ⓐ 20   Ⓑ 27   Ⓒ 30   Ⓓ 25

B Solve. Show your work.

1. In Mr. Brown's Social Studies class of 20 students, $\frac{1}{5}$ of the students scored above 90%, $\frac{1}{4}$ scored between 81 and 90, and the rest scored between 70 and 80. Find the errors in the table below and explain your answer.

Range	Students	Percent
70-80	~~IIII~~ IIII	55%
81-90	IIII	20%
90+	~~IIII~~ I	25%

2.

Dana's little sister dropped her letter blocks on the floor. If she reaches for them without looking, what is the probability of:

- picking up a B block?
- picking up an A block?

- picking up a D block?
- picking up a C block?

3. Jack found that the average yearly precipitation in Philadelphia is about 40 inches, with about 8 inches falling during July and August. He said about 20% of precipitation falls during those two months. Jack also found that about 6 inches falls during the 2 driest months, February and October. Based on this, he said there was a 15% chance any rain would fall during February and October. How do you know both of Jack's statements are true?

4. After 60 games of their season, the Boston Red Sox had won 70% of their games for the best record in the American League. In the National League, the Arizona Diamondbacks had won $\frac{3}{5}$ of their 60 games. How many games had each team won? If they continued to win games at the same rate, how many games would they win over a 162-game season?

SKILL TUNE-UP: GEOMETRY

Perimeter is the distance around a shape. Add the lengths of all the sides of a shape to find perimeter.

Area is the amount of space a shape covers. To find the area of a rectangle, multiply its width by its height. To find the area of a triangle, multiply width by height, then divide by 2.

Volume is the amount of space inside a solid object. To find the volume of a rectangular prism, multiply its width, height, and length.

When drawing **to scale**, use proportions. For example, if you are drawing a picture of your aquarium, you may not want to draw it using the actual size. Assign a smaller unit to stand for a larger unit.

A 1-inch line might represent the 2-foot tall glass side (1 inch = 2 feet). If the aquarium is 3 feet wide, draw the width with a $1\frac{1}{2}$ inch line.

A **Choose the best answer.**

1. What is the perimeter of a rectangle with a width of 5 feet and a length of 9 feet?

   Ⓐ 14 feet   Ⓒ 28 feet

   Ⓑ 45 feet   Ⓓ 32 feet

   To find perimeter, add each of the sides. $5 + 5 + 9 + 9 = 28$ feet. The correct choice is C.

2. A right triangle has a 15 centimeter base and a height of 8 centimeters. What is its area?

   Ⓐ 60 square centimeters   Ⓒ 120 square centimeters

   Ⓑ 30 square centimeters   Ⓓ 31 square centimeters

3. The base of a 12-inch tall cereal box is 3 inches wide and 5 inches long. What is the volume of the cereal box?

Ⓐ 90 cubic inches   Ⓒ 20 cubic inches

Ⓑ 120 cubic inches   Ⓓ 180 cubic inches

4. A car is 15 feet long. If a scale model is made to a scale of 1 inch = 3 feet, how long will the model be?

Ⓐ 15 inches   Ⓒ 5 inches

Ⓑ 3 inches   Ⓓ 10 inches

B Write a phrase or sentence to answer the question.

1. Mateo has a desk that measures 2 feet by 5 feet. He wants to order a mat for his desk so it isn't scratched. He looks in a catalog and orders a 20 square foot mat. Explain why this mat will be too large for Mateo's desk.

2. There are two swimming pools in Cathy's town. They each have a volume of 2,400 cubic feet. The first pool is 8 feet deep and 20 feet long, so Cathy determines that the first pool is 15 feet wide. She knows the second pool is 24 feet long and 10 feet deep, so it must be 10 feet wide. Why is Cathy correct about the dimensions of the pools? Explain your answer.

3. The United States is roughly 3,000 miles from east to west. If you wanted to drive completely around the perimeter of the country, you would drive about 9,000 miles. Jim said that this means the United States must be about 6,000 miles from north to south. Is Jim correct? Explain your answer.

4. Kelly is making a scale drawing of her room, which measures 25 feet long and 15 feet wide. She has a limited amount of paper so she wants to make the map as small as possible. The first time, she decides to use a scale of 1 inch = 5 feet. Then she considers changing the scale to 2 inches = $2\frac{1}{2}$ feet. Which scale would allow her to make the smallest possible scale drawing?

STRATEGY: DRAW A DIAGRAM

Drawing a diagram can help you solve a story problem.

Juanita wants to fix up her odd-shaped room. One wall is 10 feet long, but the opposite wall is 16 feet. Another wall is 12 feet long. The opposite side goes around the closet, which is 6 feet long, and 2 feet wide. Juanita wants to put up a wallpaper border around the wall. How long should her border be?

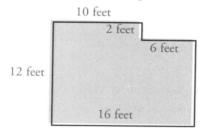

10 feet
2 feet
6 feet
12 feet
16 feet

Subtract the width of the closet from 12 to find that the length of the right side wall is 10 feet.

Finally, add the lengths of all sides.

10 + 2 + 6 + 10 + 16 + 12 = 56 feet   Juanita needs 56 feet of wallpaper for her border.

Next, Juanita needs to find the area of her room to know how much carpet to buy.

Divide the room into 2 rectangles. Then, find their area. Remember to add the areas together to find the total.

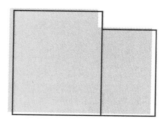

12 x 10 = 120 square feet
Total area = 180 square feet

6 x 10 = 60 square feet
Juanita needs 180 square feet of carpet

A Choose the letter of the correct answer.

1. Leon's backyard measures 40 feet by 30 feet. Along the back border, he wants to build a garden that will measure 5 feet by 8 feet. Which diagram can we use to find the area of the backyard (excluding the garden)?

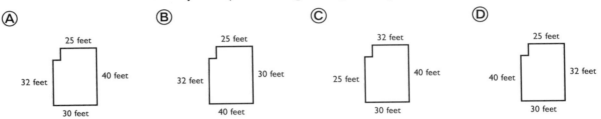

Ⓐ
25 feet
32 feet   40 feet
30 feet

Ⓑ
25 feet
32 feet   30 feet
40 feet

Ⓒ
32 feet
25 feet   40 feet
30 feet

Ⓓ
25 feet
40 feet   32 feet
30 feet

 B Solve. Show your work on another piece of paper.

1. The Tigers team wants to fix up their baseball field. They want to put in fresh dirt and grass in their infield. Their baseball diamond measures 60 feet from home plate to first base, 60 feet from first to second base, 60 feet from second base to third base, and 60 feet from third base to home plate. They plan to leave a 100 square foot area of dirt for the pitcher's mound in the middle of the diamond. When they go to buy grass seed, they find they must know the area of the diamond that will be covered in grass. Draw a diagram to help them find how much grass seed they will need.

2. The Grand Canyon is truly one of nature's wonders. Measuring 268 miles long and up to 17 miles wide, the Grand Canyon also reaches a depth of over a mile. It is not surprising that it is a "must-see" site destination for many people. One of the most popular trails in the Grand Canyon, the Bright Angel Trail goes from the rim of the canyon to the bottom, 4,600 feet below. Jackie walked the $7\frac{1}{2}$ miles from the rim to the bottom in about 5 hours. Next, she took the River Trail for another $1\frac{1}{2}$ miles to connect with the South Kabib Trail. She took this trail up to the rim, covering another $6\frac{1}{2}$ miles to the rim in 8 hours. From there, she walked back $1\frac{1}{2}$ miles to her starting point. What was the distance that Jackie hiked? Draw a diagram to help solve this problem.

STRATEGY: GUESS AND CHECK

Sometimes the easiest way to solve a problem is to make a good guess, then check to see if it is correct.

Abdul wants to wrap a gift for his friend, Cho. The box is 4 inches long and 5 inches wide. If the gift wrap he needs is 100 square inches, how tall is the box?

To find the answer, make an estimate and then check to see if it is correct.

Equation: 4 inches x 5 inches x height = 100 square inches

Guess: 3   Check: Does 4 x 5 x **3** = 100?   (Does 20 x 3 = 100?)

$60 \neq 100$

Make a better guess based on the answer above.

Guess: 6   Check: Does 4 x 5 x **6** = 100?   (Does 20 x 6 = 100?)

$120 \neq 100$

Make a better guess based on the answers above.

Guess: 5   Check: Does 4 x 5 x **5** = 100?   (Does 20 x 5 = 100?)

$100 = 100$ Abdul's gift box is 5 inches tall.

 Choose the letter of the correct answer.

1. Some workers are digging a hole. It is 3 feet wide and 5 feet long. They remove 90 cubic feet of dirt. What is a reasonable guess for the depth of the hole?

   Ⓐ 3 feet   Ⓒ 6 feet

   Ⓑ 8 feet   Ⓓ 4 feet

To find volume, multiply length x width x height(depth). Since we know the width and length, multiply 3 x 5 = 15. We also know the volume of dirt removed. 90÷15 = 6 The correct choice is C.

2. A parking lot measures 150,000 square feet and is 500 feet in length. What is the width of the parking lot?

Ⓐ 300 feet   Ⓒ 30 feet

Ⓑ 3,000 feet   Ⓓ 3 feet

3. When Jeff walks around the block he covers a distance of 500 feet. If the block is 200 feet long, what is the width?

Ⓐ 500 feet   Ⓒ 50 feet

Ⓑ 5 feet   Ⓓ 5,000 feet

4. A scale model of a building is 30 centimeters tall. What is the scale used, if the actual building is 300 meters tall?

Ⓐ 1 centimeter = 1 meter   Ⓒ 1 centimeter = 10 meters

Ⓑ 1 centimeter = 1,000 meters   Ⓓ 1 centimeter = 100 meters

B   **Solve. Show your work on another piece of paper.**

1. Mike's backyard is huge. It covers an area of 1,800 square feet, and it is twice as long as it is wide. If the backyard is 60 feet long, is 3 feet a good guess for the width? Explain your answer.

2. If you were to walk completely around the perimeter of Colorado, you would walk about 1,300 miles. The state is about 385 miles from East to West. Is 495 miles a reasonable guess of the distance from North to South? Explain your answer.

3. Mount Mansfield in Vermont is the highest peak in the Green Mountains. There are many trails to the summit at 4,393 feet. Once at the top, the views are fantastic. The shortest trail is the Bear Pond Trail, but it is also the steepest. Vicky walked up the Bear Pond Trail and then walked around on the summit and returned for the trip back down. The total distance she covered was about 6 miles. If the distance Vicky covered on the summit was about half the distance she covered walking up and down the trail, how long was the trail?

L E S S O N   4

CHOOSE A STRATEGY

Review the strategies you have learned so far.

Diagram—Drawing a diagram can help you solve a story problem.

Yakov wants to frame a triangular artwork, but he can only find square frames. There should be at least one inch space around the triangle on every side. What is the smallest frame Yakov could get to hold a 4 by 4 by 4 inch triangle (to the nearest inch)?

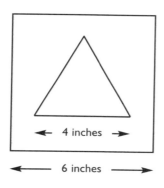

Guess and Check —Make a reasonable estimate and then check to see if it is correct.

Similar figures are proportional to one another. Which two figures below are proportionate? (Hint: Use ratios to check.)

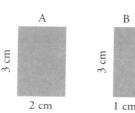

Guess: A is proportionate to C
Answer: Yes. $\frac{2}{4} = \frac{3}{6}$

  Choose the letter of the best answer.

1. A rectangle has a perimeter of 30 feet. If the width is 6 feet, what is the length of the rectangle?

   Ⓐ 18 feet   Ⓑ 90 feet   Ⓒ 19 feet   Ⓓ 9 feet

2. A box has a height of 6 inches and a length of 10 inches. If the width is half the height, what is the volume of the cube?

   Ⓐ 180 cubic inches   Ⓒ 270 cubic inches

   Ⓑ 90 cubic inches   Ⓓ 120 cubic inches

 Write a phrase or sentence to answer the question.

1. A scale model of a ship is 12 centimeters long. The scale of the model is 1 centimeter = 6 meters. Denise said that the actual ship was 2 meters long. Was Denise correct? If not, what was the actual length? Explain your answer.

2. A park has a width of 500 feet and a length that is twice as large. Draw a diagram of the park and tell how to find the perimeter.

C   **Solve. Show your work.**

1. What is the area of the figure below? Show your work.

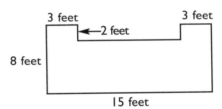

2. If a room measures 30 feet long and 25 feet wide and has a volume of 7,500 cubic feet, how high is the room? Show your work.

3. It is about 200 miles from New York to Boston. If you were asked to draw a map showing these two cities, what scale would you use? How far apart would the two cities be on the map?

L E S S O N **5**

MULTI-STEP STORY PROBLEMS

A **Choose the letter of the best answer.**

1. Which field has an area of 77 square feet?

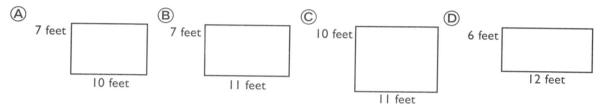

Ⓐ 7 feet, 10 feet   Ⓑ 7 feet, 11 feet   Ⓒ 10 feet, 11 feet   Ⓓ 6 feet, 12 feet

2. A rectangular box has a volume of 36 cubic meters. What are its possible dimensions?

   Ⓐ 6 meters x 6 meters x 2 meters

   Ⓑ 2 meters x 6 meters x 2 meters

   Ⓒ 3 meters x 6 meters x 2 meters

   Ⓓ 3 meters x 6 meters x 3 meters

3. A cube has a volume of 64 cubic inches. What is the length of its side?

   Ⓐ 4 inches   Ⓑ 6 inches   Ⓒ 8 inches   Ⓓ 10 inches

4. Manuel is making scale drawings of a store that is 60 feet long. If one scale drawing uses a scale of 1 inch = 3 feet and another scale drawing uses a scale of 1/2 inch = 2 feet, what is the length of each of the drawings?

   Ⓐ 20 inches and 30 inches

   Ⓑ 15 inches and 40 inches

   Ⓒ 60 inches and 20 inches

   Ⓓ 20 inches and 15 inches

Write a sentence or two to answer the question.

1. John sketched the diagram below. He used it to find that his first floor measures 1,700 square feet. Explain how John might have used the diagram to find the area Explain why John's diagram is correct.

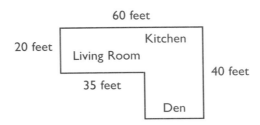

2. There are many interesting sights to see in our nation's capital Washington, D.C. One of the most popular tourist destinations is the Washington Monument. Most people choose to get to the top by elevator. Those with more energy can climb the steps. Once at the top, they have a great view of the city. Gabriela saw a model of the monument that was built to a scale of 2 inches = 100 feet. Seeing that the model was 11 inches tall, Gabriela calculated she would have an easy time walking up 50 feet of stairs. Why was Gabriela wrong?

C **Solve the following problems. Show your work.**

1. The O'Neal family is putting a fence around their property. The property is 120 feet long and has a perimeter of 360 feet. It is shaped like a rectangle. How wide is the property? If the fence sections are each 6 feet long and cost $30 per section, how many sections will they need? How much will the O'Neals spend on their fence?

2. On May 18, 1980, Mt. St. Helens erupted. Some trees as tall as 180 feet were flattened by the blast. Beside the lava and ash that was released, there was an avalanche that swept down the valley of the Toutle River. The avalanche covered 24 square miles under 150 feet of mud and debris. What are some possible dimensions of the land that the avalanche debris covered?

3. After nine hours of eruptions, a wall of lava raced down the mountain at up to 100 miles per hour. This material was very hot, almost 1,000 degrees Fahrenheit. It was so hot, that two weeks later the lava was still 780 degrees! If the lava covered an area measuring 32 square yards and had a volume of 160 cubic yards, how high was the lava flow?